CAMBRIDGE COUNTY GEOGRAPHIES

General Editor: F. H. H. Guillemard, M.A., M.D.

MIDDLESEX

Cambridge County Geographies

MIDDLESEX

by

G. F. BOSWORTH, F.R.G.S.

With Maps, Diagrams and Illustrations

Cambridge:
at the University Press
1913

CAMBRIDGE UNIVERSITY PRESS
Cambridge, New York, Melbourne, Madrid, Cape Town,
Singapore, São Paulo, Delhi, Mexico City

Cambridge University Press
The Edinburgh Building, Cambridge CB2 8RU, UK

Published in the United States of America by Cambridge University Press, New York

www.cambridge.org
Information on this title: www.cambridge.org/9781107652910

First published 1913
First paperback edition 2013

A catalogue record for this publication is available from the British Library

ISBN 978-1-107-65291-0 Paperback

CONTENTS

ILLUSTRATIONS

MAPS

The illustrations on pp. 9, 15, 23, 25, 74, 94, 96, 109, 128, 151, and 157 are from photos by Messrs F. Frith & Co.; those on pp. 17, 18, 19, 24, 26, 33, 48, 105, 106, 108, 118, 123, 135, 137, 150, and 156 from photos by Messrs J. Valentine & Sons; those on pp. 13, 28, 30, 31, 62, 65, 67, 72, 75, 97, 98, 112, 113, and 122 from photos by Mr R. B. Fleming; that on p. 71 is from a photo kindly supplied by Messrs J. I. Thornycroft & Co., Ltd.; that on p. 101 is reproduced from Mr J. S. M. Ward's *Brasses* in the series of *Cambridge Manuals*; that on p. 115 is from a photo supplied by the Homeland Association; those on pp. 139 and 141 are from portraits in the National Portrait Gallery; that on p. 81 is from a photo supplied by The London Stereoscopic Company; that on p. 129 is from a block kindly supplied by Mr R. W. B. Buckland; that on p. 154 is from a photo by Mr O. W. Dumaresq, kindly supplied by Mr A. T. Vardy.

ERRATA

p. 20, l. 11, and p. 158, l. 24. *For* the first lock *read* the first lock bearing a Middlesex name.

p. 70, l. 15, and p. 145, l. 23. Messrs Thornycroft's Works have now been removed from Chiswick.

p. 142, l. 9. *For* Dr Parkin *read* Sir William Perkin.

p. 144, l. 6 from bottom. *For* Percy *read* John.

p. 160, l. 16. *For* 129 *read* 120.

I. County and Shire. The word *Middlesex*. Its Origin and Meaning.

Middlesex is the metropolitan county of England, and only Rutland and London are smaller in point of area. For more than one thousand years it contained London, or the greater part of London, and this fact contributed to its wealth and importance. Although the whole of London north of the Thames was taken from Middlesex in 1888, the county is at the present time one of the most populous of English counties; and we shall certainly understand much better the progress and development of England as a whole, if we first carefully study the geography and history of this small, but very important, portion of our land.

In this chapter we will first consider the meanings of the two words, *shire* and *county*, and then endeavour to trace the origin of the county of Middlesex and the meaning of its name. Since the ninth century, and perhaps from an even earlier time, the county, or shire, in England has been the chief unit of local government, in much the same way that the department is regarded in France, the canton in Switzerland, and the state in

America. Although we now have the two words, shire and county, it is well to remember that before the Norman Conquest the word shire only was used.

We find that the word shire, in the earliest period of our history, simply meant a division, and was thus used to denote the various portions of Cornwall, and the two kingdoms of Kent. As time passed on, however, the word acquired a new meaning, and was applied to any portion that was *shorn* off or cut off from a larger division. The portion cut off was a *share* or *shire*, and hence many of our counties have retained this affix since the settlement of the English in our land.

The word county is of later date and is due to the Norman invaders, who identified the old English *shire* with their own *comitatus*, the district of a *comes* or *count*. And thus it comes about that we use the two words shire and county to denote the larger divisions of our land that were made long ages ago.

The counties of England differ considerably in their origins, and their names are often a key to the right knowledge of their historical development. Such counties as Middlesex, Essex, Kent, and Sussex had a different origin from Nottinghamshire, Leicestershire, and Northamptonshire. While the former are probably survivals of early kingdoms, the latter are undoubtedly shares, or shires, of the once important kingdom of Mercia. We shall see in the third chapter that, for more than a thousand years, Middlesex kept its name and boundaries, and it was not till 1888 that a great change was made in its extent. It is the knowledge of such facts as these that makes the study of county history and geography so interesting, for here we find an important clue to the continuity of our history, and to the growth and development of our local institutions.

It is rather difficult to trace the origin of some of our counties and the meaning of their names, but there is not much difficulty with regard to Middlesex. Here we have a county with a distinctly English name, which was derived from the Saxons who dwelt in this part of England. The Saxons were perhaps the strongest of the various invaders of our country in the fifth century, and they were distinguished by the kingdoms they formed along our eastern and southern shores. Thus we read of the East Saxons, South Saxons, West Saxons, and so on, while the other invaders, the Angles and Jutes, settled elsewhere. There is no evidence that Middlesex was originally a separate kingdom, and we may say with a considerable amount of certainty that it formed part of the kingdom of Essex, which at its first settlement included the modern county of that name, the present Middlesex, and possibly the whole of Hertfordshire. The name of Middlesex seems not to be mentioned by Bede, but it occurs in a charter of 704, where it is called *provincia*, and also (with the same title) in one of Ethelbald's charters. It is mentioned again in 767, and after that date it is of fairly frequent occurrence. From our scanty knowledge of the formation of the early English kingdoms, we may come to the conclusion that Middlesex represents the territory which fell to one of the kings when the original kingdom of Essex was divided. This came about at some time in the eighth century, and so we can understand why Middlesex received its name. It was the territory between, or in the middle of, the East Saxon kingdom and the West Saxon kingdom, which were conterminous. Perhaps Middlesex was separated from the East Saxon kingdom and made into a county, owing to the growth of London, which had been the capital of Essex. There is another point of interest that deserves our notice in this chapter, for the East Saxon kingdom was also the see of the bishop of London, and so it continued till quite recent times, when, owing to the growth of population, Essex and Hertfordshire were annexed to the see of Rochester, and afterwards to that of St Albans. Middlesex, however, remains in the diocese of London, another interesting fact which links the county with the earliest period of our history.

Thus we may say that the modern county of Middlesex grew out of the East Saxon kingdom, which was formed in a district that had been settled by a Celtic tribe known as the Trinobantes, or Trinovantes, who were living in Britain when the Romans first landed in our country.

2. General Characteristics — Position and Natural Conditions.

Middlesex is a small inland county on the left bank of the Thames, in the south-east of England. It is the metropolitan county of England, and as such occupies a somewhat anomalous position, for it contained within its old borders practically the whole of London north of the Thames, which was severed from it by the Local Government Act of 1888.

As we shall find in later chapters, Middlesex has suffered from the overpowering influence of London. It is a matter of history that London was founded long before Middlesex, and from the time of King John to 1889 one of the London sheriffs acted as sheriff of Middlesex. It was not till that year that Middlesex had its own High Sheriff, and even now it can hardly be said to have its own county town. Brentford is sometimes mentioned as the capital, but the county business is practically carried on in London. The County Council sits at the Middlesex Guildhall in Westminster, and there also the Quarter Sessions are held. In another way, too, the position of Middlesex is anomalous, for it is the only county that has not its own constabulary. Here again it is dominated by London, the whole county being served by the Metropolitan Police.

We shall understand the position of Middlesex much better if we realise from the outset that it has always been considered as dependent on London, or as of some use to the metropolis. From the earliest time all the great roads ran through Middlesex to London Stone, and the great forest of the county was the hunting-ground of Londoners. Even when the forest had been cleared away, the villages that succeeded it were the sources of supply for the needs of the ever-increasing London. It is not too much to say that corporate life has been crushed out of this county,

Stone at Staines

[*Marking the western boundary of the jurisdiction of the City of London on the River Thames*]

owing to the greatness of London; not a single place has grown into real importance, nor is there outside of London a building of first-rate importance with the exception of Hampton Court Palace. The villages on the Thames, such as Teddington and Twickenham, early began to increase in size because of the convenience of their position by the river and consequent accessibility from London. It is only since the extension of the railway and tramway systems that the villages to the north and north-west of London have grown in size, and their growth has been almost entirely due to the building of houses for the use of Londoners.

Two centuries ago it was written that Middlesex "is in effect but the suburbs at large of London, replenished with the retiring houses of the gentry and citizens thereof," and to-day these words are even more true when we find the great increase in population in the suburbs bordering on London. Places like Tottenham, Edmonton, Enfield, Willesden, Hornsey, Wood Green, Acton, Chiswick, and Ealing were quite country villages a hundred, or even less years ago, while to-day they have ever-increasing populations and are almost entirely dependent on London. Thus we find large tracts of Middlesex being suburbanised, and whenever the boundaries of the county of London are re-adjusted, Middlesex is bound to be the loser, as it was in 1888. Since 1901, the chief agency at work in the development of Middlesex has been the electric tramway systems, which now run along some of the main high roads of the county in the direction of Hounslow, Uxbridge, Enfield, Edgware, Tottenham, and Finchley.

As an agricultural county Middlesex is of some importance, for about three-fifths of its area are under crops and grass. The greater portion of this acreage is used for pasturage and dairy farming, while increasing attention is being paid to fruit-growing to meet the needs of the metropolis, where the Middlesex farmers and market-gardeners find a ready market for their produce.

Middlesex has little claim to rank as an industrial county. We have only to remember that in 1911, 1,078,334 people were living in urban districts near London, while the rural population was 48,131, in order to understand how Middlesex is completely overshadowed by the importance and proximity of London. On the other hand, industries are rapidly springing up at such centres as Willesden, Acton, and Hayes, and also along the Lea valley.

Fortunately, however, there are parts of Middlesex which retain some of their old characteristics. A recent writer on this county says that, "though the great corn-land of Middlesex has largely become pasture or market-gardens, and old-fashioned farmsteads are few and far between, there are still some rustic 'bits' to be seen away from the tram-dominated highways." The same writer also remarks that "if in Middlesex we have a district lacking any of the more striking beauties even of some of its nearest neighbours, we have one that can vie with the best of them in the variety and multiplicity of its associations with men and events." There is scarcely a parish in the county without its memories of some one who made himself famous in the great metropolis, and we shall find, in a later chapter, that Middlesex is particularly rich in its associations with poets, from Milton and Pope down to Keats and Tennyson. In his *Autobiography*, Leigh Hunt prided himself on being born in the "sweet village of Southgate"; and rather more than a century ago, he wrote, "Middlesex in general is a scene

Shepperton

of trees and meadows, of 'greenery' and nestling cottages; and Southgate is a prime specimen of Middlesex."

It has been remarked that the proximity of Middlesex to London has kept it largely free of the horrors of battle, and the same association has made it in the past a place for the stately homes of princes and noblemen, of statesmen and city merchants. There is no doubt, too, that its nearness to the capital has had an important share in making its open spaces famous in the annals of highwaymen.

It is not possible to claim that Middlesex is a beautiful county in the same sense as Surrey, yet there are picturesque spots and beautiful villages. The Thames valley is delightful at Twickenham and Teddington, and in the higher reaches at Shepperton and Laleham. The views across the Colne from Harefield are typical pastoral scenes, while the stretches of undulating country in the Stanmore district are altogether pleasant. The most picturesque villages are Ruislip, Ickenham, Northolt, Harefield, and Cranford, and perhaps the most famous view is obtainable from Harrow churchyard. Here on the top of Harrow Hill, at a height of rather more than 400 feet, the view is said to extend into thirteen counties. The view from the churchyard has been made famous for all time by Byron, who wrote to his publisher, Murray, in 1822, "there is a spot in the churchyard near the footpath, on the brow of the hill, looking towards Windsor, and a tomb under a large tree where I used to sit for hours when a boy. This was my favourite spot."

3. Size. Shape. Boundaries.

We have seen in the previous chapters how Middlesex came to be a county, and we have also learnt something of its general characteristics. Now we are in a position to consider its size and shape, and with the help of a good map we may define its boundaries. At the outset, it is well to remember that the Ancient County of Middlesex was much larger than the present county; which was reduced from its former size in 1888, when the County of London was formed from Middlesex, Surrey, and Kent.

The area of the entire Ancient County, inclusive of the parishes in the County of London, was 181,320 acres, or about 283 square miles, while the area of the present county is 149,668 acres, or nearly 234 square miles. Only Rutland and London have a smaller area among the counties of England, and it occupies about one-two-hundred-and-fiftieth of the entire area of England and Wales. The extreme distance from Chertsey Bridge in the south-west to Waltham in the north-east is about 28 miles. The greatest length from north to south in the western portion is 18 miles, and in the eastern 11 miles; while the greatest measurement due east and west is 19 miles.

The county is comprised within very irregular outlines, especially on the north and south sides. Before London was taken from Middlesex, the county had the shape of a rough parallelogram. Now it may be described as a very irregular quadrilateral, having only the east and west sides at all regular.

When we deal with the boundaries of any county, we have always some interesting questions to answer. Although it is comparatively easy to state the present boundaries of a county, it is not always so easy to say how and when these boundaries were settled. However, with the aid of a good map and a knowledge of its early history, we may get a fairly accurate idea.

Middlesex is bounded on the north by Hertfordshire, and the very irregular line of division is marked by no physical features. The western portion of this northern boundary is about 18 miles long and is five miles further south than the eastern and shorter portion. A line from Mill End to Southgate would cut the western half no less than six times, although the boundary would nowhere be distant more than a mile on either side of that line. From Southgate the boundary takes a north-westerly direction returning upon itself through Chipping Barnet, so as almost to inclose the spur of Hertfordshire containing Totteridge and Barnet. About a mile to the west of Chipping Barnet, the boundary turns north for some four miles, and then proceeds almost direct to Waltham Cross in the east. Now we may ask the question, How can we account for this very irregular northern boundary? In attempting an answer, history comes to our aid, and we may conclude that the northern boundary is partly a line obviously parallel to whatever remains of an ancient earthwork, known as Grims Dyke, and partly an extremely irregular boundary determined by the estates which belonged to the Abbey of St Albans.

The boundary on the east is easy to determine, for from Waltham Abbey in the north to Tottenham in the south the river Lea divides the county from Essex; and the western boundary is the river Colne, which separates Middlesex from Buckingham from Maple Cross to Staines, where that river joins the Thames. Here then we have two geographical boundaries, which were the original lines of separation of the kingdom of the Middle Saxons from the East Saxons and the West Saxons.

Before the separation of London from Middlesex in 1888, the Thames formed the entire southern boundary from Staines in the west to Blackwall on the east, where the river Lea joins the Thames. Now the Thames forms

The Lea at Clapton

this boundary only from Staines to Chiswick, and the new Middlesex and London boundary begins at Hammersmith, and proceeds irregularly in a generally northeasterly direction to the Lea.

We may conclude our investigation of the boundaries of Middlesex by stating that this county was originally a buffer state between the two powerful east and west Saxon kingdoms, and that its area was not appreciably affected till the formation of the County of London in 1888, when it lost about 50 square miles of its former area.

4. Surface and General Features—Forests, Commons, and Open Spaces.

Middlesex differs from the other divisions of Metropolitan England inasmuch as it is wholly removed from the sea, and is quite devoid of any definite orographical features. A careful study of the map will show that Middlesex, if we include London, is almost completely surrounded by rivers so as to be converted into an inland island, and to be practically cut off from the physical system of the adjoining counties.

The general surface of Middlesex is a plain sloping gradually southwards to the Thames. The whole district in the south-west is almost a dead level, for it is nowhere more than 20 feet and often less than 10 feet above the Thames at Staines Bridge. The northern portion of the county has the highest hills, but these are only high in comparison with the lowland along the Thames, the Colne, and the Lea. Near the Hertfordshire boundary are the eminences of Deacon's Hill, Brockley Hill, and Elstree Hill, and from these the high grounds extend east to Highwood Hill and Mill Hill, and west to Stanmore Common, Harrow Weald Heath, and Pinner Hill. The highest point of these northern hills and indeed in the whole

Hampstead from Parliament Hill

county is at the county boundary stone at Hart's Bourn on the Bushey and Watford Road, where a height of 503 feet is reached. The other hills mentioned are from 400 to 450 feet in height.

There is another shorter and lower range to the north-east of London, which is often called the "Northern Heights" of London. Hampstead and Highgate are the best-known hills and attain a height of 442 feet and 425 feet respectively. The north-east of Middlesex is a wedge of fairly high ground between Enfield and Barnet, sloping down to Wood Green in the south, while to the east of this is a broad belt of level land in the valley of the Lea. Besides the heights already mentioned, there is no other considerable eminence in the county except the isolated hill of 405 feet on which Harrow church stands.

Middlesex was once covered with a great forest that extended north from Hampstead and Highgate, and only one hundred years ago the waste and common lands in this county were about 29,000 acres in extent. Most of the woodlands, the heaths, and the greens are now only remembered by name, while nearly all the larger commons have been enclosed, and such extensive tracts as Finchley Common, Hounslow Heath, and Enfield Chase are no longer open spaces.

Enfield Chase was a fine tract of woodland extending over 4000 acres. It was preserved as a hunting ground, and many of our kings, especially James I, often hunted in it. An act for its disafforestation was passed in 1777, and now the Chase is practically all enclosed, the only portions left being Hadley Wood, Hadley Common, and Trent Park.

Hounslow Heath was an expanse of 5000 acres of barren and uncultivated land which lay on either side of the two famous converging roads to London from the west of England. Since 1801 much has been enclosed,

East Finchley

but enough remains to show what a dreary, flat waste it was in the early days. Hounslow Heath has many historical associations, and it was long notorious for the highwaymen who frequented this district.

Finchley Common comprised more than 2000 acres of waste and uncultivated land. It shared with Hounslow

Heath the reputation of being the favourite hunting-ground of highwaymen, and was not considered safe for travellers by the Great North Road as late as the end of the eighteenth century. This common has long ceased to exist, and Finchley Cemetery of 90 acres was taken from it some years ago.

Highgate Woods

Bushey Park, so called from the number of its bushes, is 1110 acres in extent, and is one of the royal parks of Hampton Court. It is pleasantly wooded, and its principal feature is the fine avenue of chestnuts. In the park is an artificial river known as the Cardinal's river, and there are also two small lakes in its eastern part.

In recent years several fine open spaces have been assigned to public use, and among the more noteworthy of these may be mentioned Highgate Woods and Finsbury Park.

Highgate Woods of 70 acres, a portion of the great forest of Middlesex, were presented to the public by the Corporation of the City of London in 1886.

The Lake, Finsbury Park

Finsbury Park, of 115 acres, was opened in 1869. It owes its name to the fact that it was originally intended to serve as a public park for the people of the borough of Finsbury, which then reached as far as its southern border. This park is of much value to the rapidly increasing population of the district, and there is ample provision for the recreation of visitors. The ornamental lake has an island which forms a sanctuary for birds, and the flower-beds, the rockeries, and the American garden are some of the special features of this park.

5. The River Thames.

The Thames may be considered as belonging to Middlesex from Staines to Hammersmith. Middlesex is on its left bank for about 27 miles, and the Thames is navigable the whole of this distance, although it is only tidal eastwards from Teddington. The first lock on the Thames is at Teddington, and Staines is generally regarded as the dividing point of the Upper and Lower Thames, where the authority of the City of London over the river ceases.

As the Thames is the greatest of all our rivers, it will be well to devote a little time to a consideration of its source and course before it becomes a Middlesex river. Throughout our history and literature the Thames plays a prominent part, and we shall find in the pages of this book many references to it. With our English poets it has been a favourite theme, and such expressions as "the silver-streaming Thames" are of frequent occurrence, while its distinction has been marked by the well-known name "Father Thames." It is interesting to note that the Thames, or Tamesis as it was once called, is the earliest British river mentioned in Roman history. Its name is of Celtic origin, and its derivation is probably the same as that of the Tame, the Teme, and the Tamar, in other parts of England.

The Thames rises at Coates in Gloucestershire not far from the borders of Wiltshire. By the side of a canal that runs through that little village there is an ash tree, encircled with ivy, and in its bark are cut two letters, T. H.—"Thames Head." At the foot of the tree are some stones indicating that a spring runs under the canal. Near by are other springs, and the water from them flows through the meadows and soon enters the county of Wilts as the Thames. The upper part of the main stream is often called the Isis, and not the Thames, until it has received the waters of the Thame near Dorchester in Oxfordshire. The Thames has Oxfordshire, Buckinghamshire, Middlesex, and Essex on its left bank, and Wiltshire, Berkshire, Surrey, and Kent on its right bank, while the County of London is on both banks. From its source in the Cotswold Hills to the Nore the length of the main stream is estimated to be about 220 miles, and the drainage area of the Thames is reckoned to be about 5244 square miles.

Although water is received from springs that issue in the Cotswold Hills at heights of about 800 feet, the main supplies are derived from levels below 400 feet. Along the higher courses the fall of the stream is nearly 7 feet per mile, but from Teddington to London Bridge it is rather less than 1 foot per mile. The range of the tides at London Bridge is about 16 or 17 feet, and at Twickenham 8 feet. The low-water depth of the river at London Bridge is 9 feet, while at Brentford and Teddington it is from 4 to 6 feet. The tides, which at London Bridge ebb seven hours and flow five hours, extend to the weir at Teddington, where the average daily flow of the river is estimated at 1110 millions of gallons. A reference to the map will show that the Thames has a very winding course, and between its source and Teddington there are numerous locks, or water-gates.

The river reaches the county of Middlesex at Staines, a town of some antiquity. A stone by the river-side is known as London Stone, and marks the extreme limit of the authority of the City of London over the Thames. The stone, which is much decayed, bears the inscription, "God preserve the City of London," and the names of the Lord Mayor who set it on its pedestal, and others who visited it in their official capacity (see p. 6). On the base is written, "Conservators of the River Thames, 1857." The bridge across the river at Staines is one of the most important as well as one of the most ancient bridges crossing the Thames. It is a handsome stone structure, designed by Rennie, and was opened by King William IV and Queen Adelaide in 1832. Two miles from Staines Bridge is Penton Hook Lock, the entrance to a long bend of the river, originally called "Penty Hook," a spot in high favour with anglers.

From Staines the course of the river is south-east; and at a distance of two miles the pretty little village of Laleham is reached. Here it was that Dr Arnold lived and worked before his appointment to the Head

Teddington Lock

Mastership of Rugby, and his son, Matthew Arnold, finds a resting-place in the beautifully-kept churchyard. Beyond Laleham is Shepperton Lock, where the average fall of the river is 6 feet 4 inches. Facing Shepperton Lock is a small island in the river, and we shall find at various spots between Staines and Chiswick that small islands of a muddy or gravelly nature occur along the Thames.

Staines Bridge

They are known as aits or eyots, and in some cases are quite a feature in the river's course. Shepperton is a river-side village about three-quarters of a mile below the lock and is a noted resort for anglers. The "Deeps" have yielded many a fine salmon in the old days, and still contain trout, barbel, perch, and jack. Halliford is only a short distance from Shepperton, lying at a most picturesque bend of the river, and near by is Cowey Stakes, where Julius Caesar is said to have crossed the river, 54 B.C. A number of stakes have been taken from this spot; but their antiquity and purpose are matters of dispute. After passing through the lock at Sunbury, we reach Hampton in two miles. A little below Hampton is an island, behind which is a beautiful backwater, where

Laleham

are moored some of the most elaborately decorated house-boats on the river. At Hampton Court an iron bridge crosses the river, while Hampton Wick has a long frontage to the Thames extending from Kingston Bridge to Teddington. There are here many fine houses which have their lawns reaching to the water's edge.

We soon reach Teddington, and notice that there is

one lock for pleasure-boats, and another, recently constructed, for barges. Teddington is a favourite haunt of anglers, and with boating men its celebrity is due to its possessing the first lock on the Thames from its mouth. Twickenham is the next village of importance on the

Eel Pie Island

Middlesex shore, and Twickenham Eyot, better known as Eel Pie Island, has been a favourite resort of pic-nic parties for three centuries. Isleworth is a place of considerable antiquity between Twickenham and Brentford, and opposite the Old Deer Park at Richmond. Brentford derives its name from the river Brent, which here falls into the Thames through locks after receiving the waters of the Grand Junction Canal. The stone bridge over the Thames which connects Brentford with Kew was opened by King Edward VII in 1902, and took the place of a bridge dating from 1789. Chiswick, in a great loop of the Thames, to the south of the main road from Brentford to Hammersmith, lies in a flat and low district largely devoted to market gardens. The most interesting part of the river-front is known as the Mall, and extends from Chiswick Church to near Hammersmith Church. The Thames ceases to be a Middlesex river just beyond Chiswick Eyot, and from that spot it belongs to the County of London.

6. The Tributaries of the Thames—Colne, Yeading Brook, Crane, Brent, Little Ealing Brook, Stanford Brook, Lea. The New River.

Middlesex is a well-watered county, and all its rivers which flow into the Thames have a general southerly course. The Colne on the west, and the Lea on the east, are the most considerable streams, and form the boundary on either side. Between these two rivers, from west to east, the other tributaries of the Thames are the Yeading Brook and Crane, the Brent, with Dollis Brook, the Little Ealing Brook, and the Stanford Brook. It would hardly be right to speak of the high land in the north of Middlesex as the watershed, although the Crane and the Brent have their sources in that neighbourhood. Both the longer rivers rise outside the county and the water-parting is further north, but the whole of Middlesex belongs to the Thames drainage area.

The Colne, formerly known as the Ux, drains a large area of chalk beyond Middlesex, and rises in the neigh-

The Colne at Uxbridge

bourhood of Tittenhanger, between St Albans and Hatfield in Hertfordshire. It forms the county boundary from a point between Harefield and Rickmansworth to Staines, where it enters the Thames after a course of about 35 miles. On the left bank it receives the Ruislip Brook, which rises in the high ground at Stanmore, and, flowing through Pinner, joins the main stream at Yiewsley.

At West Drayton the river divides; the Colne keeping to the east, the Wraysbury Stream taking an intermediate course, and the Colne Brook flowing on the west. The "crystal Colne" is often alluded to by some of our poets: Milton, who knew the Colne well, for he lived at Horton, near one of the western branches, has a pleasing reference to it in one of his Latin poems; and the river-poet writes:—

> "most transparent Colne
> Feels with excessive joy her amorous bosom swoln,"

evidently referring to its junction with the Thames. The country through which this river flows is very pretty, and here and there its water is the motive power for many mills. Much of the flat lower valley of the Colne reminds one again and again of the East Anglian lowlands. Its farm-houses, its streams, its willows, its broad pools with their reeds, and its flat pastures with abundant sheep, transport us in fancy to the scenes by the Stour or other Suffolk and Essex rivers.

The Yeading Brook and Crane form one stream, which rises on the north and west of Harrow, flowing past Yeading, Cranford, and Twickenham into the Thames by two branches, one at Isleworth, and the other about half a mile up stream. The left branch is said to have been cut by the monks of Syon to convey water-power to their flour-mill. The lower stretch of the Crane from between Twickenham and Whitton, for a distance of nearly three miles, is closely preserved as a trout stream, and affords excellent sport to the angler. The Crane drains an area of about 65 square miles, and along its banks are numerous powder, snuff, and paper mills.

The Brent, with Dollis Brook, rises in the Mill Hill district and in the north-western part of Hampstead Heath. It runs a course of about 18 miles, and drains an area of about 65 square miles. It passes Finchley and

The Brent, from Hanwell
(*Showing the bridge carrying the Uxbridge road*)

Hendon, and then supplies the Brent or Kingsbury Reservoir, a large sheet of water covering some 350 acres. This reservoir was opened in 1839 for the supply of the locks of the Paddington and Grand Junction Canal. It is well stocked with fish, and is the chief resort of water birds in the vicinity of London. In 1841 the dam burst, and caused great destruction through the Brent valley. After leaving the reservoir, the Brent has a winding course through Perivale and Greenford to Brentford, where its waters are impounded by a lock and mingle with the Thames opposite Kew Gardens. The Brent is a pleasant stream in its upper portion, but in the lower

Kingsbury Reservoir

part of its course it is liable to floods after heavy rains. The origin of the word *Brent* is somewhat uncertain, but there is no doubt that it gives its name to Brentford.

The Little Ealing Brook drains a small area south of Ealing, and joins the Thames east of Brentford. The Stanford Brook, now mostly buried, flowed through Acton Vale to the Thames at Hammersmith.

The Lea, or Lee as it is sometimes written, is the boundary between Middlesex and Essex from Waltham Cross to Tottenham, and from the latter town to Blackwall it belongs to the County of London on the right bank. The name of the river is of English origin; and although the Lea is not now of much importance as a commercial highway, it has played an important part in the history of our land from the time of Alfred. Michael Drayton sang its charms, and Izaak Walton walked by its banks, fished in its waters, and

> "Found the longest summer day too short:
> To thy loved pastime given by sedgy Lea."

The Lea has its head waters at Houghton Regis, near Dunstable in Bedfordshire, and at Seagrave not far from Luton in Bedfordshire. In its upper portion the Lea has considerable charm, especially for anglers; but since a new channel, or the New Navigation has been cut from Enfield Locks to Hackney, the lower portion has been ruined as a fishing stream. This section of the Lea is now marked by a series of reservoirs belonging to the Metropolitan Water Board, which supply water to about two million people. The Lea has one little tributary in Middlesex known as Pymmes Brook. It rises in Hadley Woods, passes Southgate, and then turns east to Edmonton. Here it is known as the Wash and finds its way into the Lea.

In this chapter mention may be made of the New River, or Myddelton's Water as it was also called. It commences at the springs of Amwell and Chadwell in

Hertfordshire, and after a winding and pretty course of about 30 miles ends at Islington. The New River was constructed in the reign of James I by a Welshman, Hugh Myddelton. He was a mine-owner and goldsmith, and a man of great energy. This waterway took five years to form, and cost him half a million of money. In

The New River, Clissold Park

addition to the engineering difficulties, Myddelton had strong opposition to his project, and it was only by King James I coming to his help that he was enabled to bring his work to a successful end in 1613. The New River runs parallel to the Lea, from which it draws much of its water, besides the supply from a number of wells. The Chadwell spring yields from 2 to 3½ million gallons of water daily. Elm-wood water-pipes, which were at first used, are occasionally dug up in some of the older streets; but they were replaced early in the nineteenth century by iron pipes.

7. Geology.

By Geology we mean the study of the rocks, and we must at the outset explain that the term *rock* is used by the geologist without any reference to the hardness or compactness of the material to which the name is applied; thus he speaks of loose sand as a rock equally with a hard substance like granite.

Rocks are of two kinds, (1) those laid down mostly under water, (2) those due to the action of fire.

The first kind may be compared to sheets of paper one over the other. These sheets are called *beds*, and such beds are usually formed of sand (often containing pebbles), mud or clay, and limestone, or mixtures of these materials. They are laid down as flat or nearly flat sheets, but may afterwards be tilted as the result of movement of the earth's crust, just as you may tilt sheets of paper, folding them into arches and troughs, by pressing them at either end. Again, we may find the tops of the folds so produced worn away as the result of the action of rivers, glaciers, and sea-waves upon them, as you might cut off the tops of the folds of the paper with a pair of shears. This has happened with the ancient beds forming parts of the earth's crust, and we therefore often find them tilted, with the upper parts removed.

The other kinds of rocks are known as igneous rocks; they have been melted under the action of heat and become solid on cooling. When in the molten state they have been poured out at the surface as the lava of volcanoes, or have been forced into other rocks and cooled in the cracks and other places of weakness. Much material is also thrown out of volcanoes as volcanic ash and dust, and is piled up on the sides of the volcano. Such ashy material may be arranged in beds, so that it partakes to some extent of the qualities of the two great rock groups.

The relations of such beds are of great importance to geologists, for by means of these beds we can classify the rocks according to age. If we take two sheets of paper, and lay one on the top of the other on a table, the upper one has been laid down after the other. Similarly with two beds, the upper is the newer, and the newer will remain on the top after earth-movements, save in very exceptional cases which need not be regarded here, and for general purposes we may look upon any bed or set of beds resting on any other in our own country as being the newer bed or set.

The movements which affect beds may occur at different times. One set of beds may be laid down flat, then thrown into folds by movement, the tops of the beds worn off, and another set of beds laid down upon the worn surface of the older beds, the edges of which will abut against the oldest of the new set of flatly deposited beds, which latter may in turn undergo disturbance and renewal of their upper portions.

Again, after the formation of the beds many changes may occur in them. They may become hardened, pebble-beds being changed into conglomerates, sands into sandstones, muds and clays into mudstones and shales, soft deposits of lime into limestone, and loose volcanic ashes into exceedingly hard rocks. They may also become cracked, and the cracks are often very regular, running in two directions at right angles one to the other. Such cracks are known as *joints*, and the joints are very important in affecting the physical geography of a district. Then, as the result of great pressure applied sideways, the rocks may be so changed that they can be split into thin slabs, which usually, though not necessarily, split along planes standing at high angles to the horizontal. Rocks affected in this way are known as *slates*.

If we could flatten out all the beds of England, and arrange them one over the other and bore a shaft through them, we should see them on the sides of the shaft, the newest appearing at the top and the oldest at the bottom, as in the annexed table. Such a shaft would have a depth of between 10,000 and 20,000 feet. The strata beds are divided into three great groups called Primary or Palaeozoic, Secondary or Mesozoic, and Tertiary or Cainozoic, and the lowest of the Primary rocks are the oldest rocks of Britain, and form as it were the foundation stones on which the other rocks rest. These are spoken of as the Pre-Cambrian rocks. The three great groups are divided into minor divisions known as systems. The names of these systems are arranged in order in the table, and the general characters of the rocks of each system are also stated.

	Names of Systems	Subdivisions	Characters of Rock
TERTIARY	Recent Pleistocene	Metal Age Deposits Neolithic ,, Palaeolithic ,, Glacial ,,	Superficial Deposits
	Pliocene	Cromer Series Weybourne Crag Chillesford and Norwich Crags Red and Walton Crags Coralline Crag	Sands chiefly
	Miocene	Absent from Britain	
	Eocene	Fluviomarine Beds of Hampshire Bagshot Beds London Clay Oldhaven Beds, Woolwich and Reading [Groups Thanet Sands	Clays and Sands chiefly
SECONDARY	Cretaceous	Chalk Upper Greensand and Gault Lower Greensand Weald Clay Hastings Sands	Chalk at top Sandstones, Mud and Clays below
	Jurassic	Purbeck Beds Portland Beds Kimmeridge Clay Corallian Beds Oxford Clay and Kellaways Rock Cornbrash Forest Marble Great Oolite with Stonesfield Slate Inferior Oolite Lias—Upper, Middle, and Lower	Shales, Sandstones and Oolitic Limestones
	Triassic	Rhaetic Keuper Marls Keuper Sandstone Upper Bunter Sandstone Bunter Pebble Beds Lower Bunter Sandstone	Red Sandstones and Marls, Gypsum and Salt
PRIMARY	Permian	Magnesian Limestone and Sandstone Marl Slate Lower Permian Sandstone	Red Sandstones and Magnesian Limestone
	Carboniferous	Coal Measures Millstone Grit Mountain Limestone Basal Carboniferous Rocks	Sandstones, Shales and Coals at top Sandstones in middle Limestone and Shales below
	Devonian	Upper, Mid, Lower — Devonian and Old Red Sandstone	Red Sandstones, Shales, Slates and Limestones
	Silurian	Ludlow Beds Wenlock Beds Llandovery Beds	Sandstones, Shales and Thin Limestones
	Ordovician	Caradoc Beds Llandeilo Beds Arenig Beds	Shales, Slates, Sandstones and Thin Limestones
	Cambrian	Tremadoc Slates Lingula Flags Menevian Beds Harlech Grits and Llanberis Slates	Slates and Sandstones
	Pre-Cambrian	No definite classification yet made	Sandstones, Slates and Volcanic Rocks

With these preliminary remarks we may now proceed to a brief account of the geology of the county.

The geology of Middlesex is better known than that of any other English county, and the reason for this is not far to seek when we remember that this metropolitan county is the residence of many workers in science, some of whom have given much time and thought to the geological records of Middlesex. Then, too, London is the headquarters of various geological and geographical societies, and in its great museums are complete illustrations of all that is connected with the records of the past as far as geology is concerned.

A reference to the geological map of Middlesex will show the extent of the strata or geological formations which occur immediately beneath the soil. In this area the strata consist of gravel, sand, brick-earth, clay, marl, limestone, and peat, and their order of succession has been proved by observation in pits and quarries, and by the records of wells and borings.

When writing the history of a country, the historian begins with the earliest records of its people, and following this plan in describing the geology of Middlesex we will begin with the Chalk formation, which forms the foundation of the entire area. This formation of soft white limestone with flints attains a thickness of about 700 feet, but owing to various causes it is subject to considerable variation. Little of the surface of Middlesex is occupied by the Chalk, but it enters the county on the east side of the Colne Valley near Harefield, and again on the north-west side of South Mimms. The Chalk may be reached anywhere in Middlesex at depths of 400 or 500 feet, or even less, and the whole of the Chalk has been passed through in a deep boring at Park Royal, Willesden. The area included in Middlesex takes in only a small portion of the Chalk which forms the London Basin. The southern margin is to be found in the North Downs, and the northern margin occurs in the Chiltern Hills; and there is thus every reason to believe that the beds of Chalk between these hills are continuous beneath the surface.

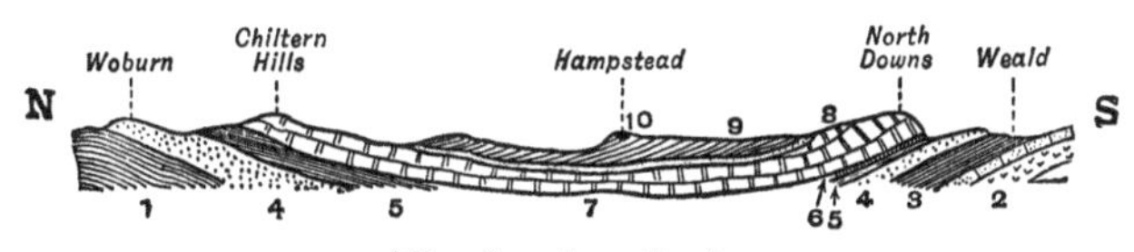

The London Basin

1. Oxford Clay.
2. Hastings Beds.
3. Weald Clay.
4. Lower Greensand.
5. Gault.
6. Upper Greensand.
7. Chalk.
8. Lower London Tertiaries.
9. London Clay.
10. Bagshot Beds.

Above the Chalk are the Woolwich and Reading Beds, which comprise plastic clays and mottled sands. The clays are for the most part brightly coloured, while the sands are white, grey, and often deep red. The Reading Beds have been traced from a point two miles north of Uxbridge, by Harefield, and again near South Mimms. The same strata are exposed at Pinner, Ruislip, and Ruislip Wood, where the junction with the Chalk may be seen in several places. Sections are from time to time opened in the neighbourhood of Pinner and Northwood for foundations and drainage, and mottled clay has been observed near the surface in places between Pinner and Wealdstone. The Reading Beds have been exposed at the Woodcock Hill Kiln, between Harefield and Rickmansworth; and a complete section has been opened up at the Harefield cement works.

We now come to the London Clay which covers the greater part of the surface of Middlesex, which has been called the county of one formation, for in no other county does a single rock formation constitute so large a proportion of the surface. The mass of the London Clay when exposed at the surface is a stiff brown clay, in which occur layers of septaria, or rounded nodules of impure carbonate of lime, which often enclose fossils. At the surface the colour of the clay is brown, for the carbonate of iron, which tinges the lower beds blue, changes to brown when exposed to the action of the atmosphere. The upper portion of the formation is somewhat sandy, and good bricks are made from it. The thickness of the London Clay increases from west to east, from 330 to about 450 feet. The London Clay is exposed in some of the deeper brick-pits; and in the neighbourhood of Highgate and Hampstead many fossils have been collected in this formation. These fossils indicate a warm climate, for they include fossil turtles, fruits of palms, and such shells as Nautilus, Voluta, etc. The area of the London Clay is undulating and picturesque, with well-timbered hedgerows of oak, elm, and ash. On rising ground, with good natural drainage, the London Clay sites for building are often better than those on shallow gravel in lower grounds, where the porous strata are liable to be water-logged.

The strata immediately succeeding the London Clay comprise a series of sands, clays, and pebble beds known as the Bagshot Beds. This formation occupies some of the higher grounds, including the prominent hills of Harrow and Highgate, which form such marked features in the Middlesex landscape. The full thickness of the Bagshot Beds is not more than 120 feet, and this formation is almost, if not entirely, without fossils.

Above the Bagshot Beds are some deposits of doubtful age. They comprise beds of flint gravels and occur in patches at considerable heights. Their age classes them as newer than Bagshot Beds, and older than the Glacial Beds. These deposits of doubtful age are well seen at Stanmore Common, where at a height of 504 feet we find them to consist of well-rounded and weathered pebbles, with some small white quartz pebbles. Sections of the gravel are to be seen here and there, including many shallow pits on Harrow Weald Common on the western side of the hill. It may be noted that these and similar gravels are variously named by geologists as Southern Drifts or Pebbly Gravels.

These deposits are succeeded by the Glacial Drift, and this formation is more complicated and difficult to map and measure than the older formations. It seems probable that after the deposit of the Bagshot Beds, the surface of Middlesex continued as dry land for many ages, for the surface of the older rocks shows evidences of long exposure to the action of the atmosphere. The

Glacial Drift rests upon the stratified rocks, and they contain, in Middlesex, no fossils. Some of the included rock fragments are polished and scratched in a way that may be accounted for by the action of ice. The earliest of these deposits are formed of sand and gravel, occasionally having masses of clay interspersed. They stretch south to Finchley, and may be seen at Hendon, Southgate, Colney Hatch, and Muswell Hill. Above them comes the Chalky Boulder Clay, which may be a deposit left by a great glacier which stretched as far south as the northern edge of the Thames valley. This is a dark bluish clay, with fragments of all kinds of rocks, besides fossils rubbed out of older rocks. It is exposed in brick-yards at Finchley, Southgate, Enfield, and Potter's Bar.

Later in age than the Glacial Drift are the Post Glacial Deposits, or the River Drift as they may be called. They were formed by the rivers when they ran at a much higher elevation than they do at present. After the Glacial Period, the rivers began their work as agents of erosion, and for thousands of years they carried down immense quantities of matter to the North Sea. Then England was part of Europe, and the Thames ran across the plain which now forms the bed of the North Sea to join the Rhine. In the beds of gravel, sand, and brick-earth then deposited, the remains are found of such huge animals as the mammoth, rhinoceros, and hippopotamus, with the hyena, wolf, lion, and bear, which crossed over when as yet the Strait of Dover had no existence. With the remains of these animals we find also the first traces of man with various forms of rude palaeolithic flint implements. The brick-earths of this period may be seen in the pits at West Drayton, Southall, Hanwell, and Ponder's End.

Our survey of the geology of Middlesex must end with a short reference to the Alluvium, a term applied to the latest deposits which form the flat meadow and marsh lands that border the Thames and its tributaries. They comprise lands now liable to flood when the rivers overflow their banks, and the more conspicuous tracts of Alluvium are along the Colne Valley from Watford to Uxbridge and Staines. There is a stretch of Alluvium between Chiswick and Hammersmith, and along the Lea valley there are broad tracts of marsh land from Enfield to Hackney and further south. The Alluvium consists of deposits of mud and loam, with occasional seams of gravel, which border the present streams, and have evidently been formed by their overflowing in comparatively recent times.

8. Natural History.

Various facts, which can only be shortly mentioned here, go to show that the British Isles have not existed as such, and separated from the Continent, for any great length of geological time. Around our coasts, for instance, are in several places remains of forests now sunk beneath the sea, and only to be seen at extreme low water. Between England and the Continent the sea is very shallow, but a little west of Ireland we soon come to very deep soundings. Great Britain and Ireland were thus originally part of the Continent, and are examples of what geologists call continental islands.

But we also have no less certain proof that at some anterior period they were almost entirely submerged. The fauna and flora thus being destroyed, the land would have to be re-stocked with animals and plants from the Continent when union again took place, the influx of course coming from the east and south. As however it was not long before separation occurred, not all the continental species could establish themselves. We should thus expect to find that the parts in the neighbourhood of the Continent were richer in species, and those furthest off poorest, and this proves to be the case both in plants and animals. While Britain has fewer species than France or Belgium, Ireland has still less than Britain.

For all practical purposes the flora and fauna of Middlesex may be regarded as more or less identical with those of the south-east of England; but of course there are certain local differences which are accounted for by differences of elevation, soil, geological formation, climate, and the presence or absence of forest. Compared with the adjoining county of Essex, Middlesex is much smaller and has no sea coast, and both its greater size and the neighbourhood of the sea must give Essex an advantage with regard to the number and variety of its species. However, although Middlesex is relatively small, there is considerable interest in the study of its flora and fauna. Some years ago when a *Flora* of this county was compiled, it was estimated that there were 859 species, of which

768 were native and 91 were introduced, and more or less completely naturalised. Owing to the rapid growth of population, and the disappearance of parks, fields, and open spaces, there is no doubt that the number of species is now much less.

The lichens of Middlesex are the poorest of any county in England, the dense and smoky atmosphere surrounding the metropolis being unfavourable to their growth. It is only in the higher grounds of the county, in the north and west and at some distance from London, that the more frequent and universally distributed species are found.

The neighbourhood of Harefield, Harrow, Highgate, and the valleys of the Colne, the Brent, and the Lea are comparatively rich in their flora. Traveller's joy and meadow rue are rather rare, but may be found near Harefield, Whetstone, and Greenford. The sweet violet is rather common in hedgerows and shrubberies about Harrow and Harefield. The water violet is frequent at Harefield and on the Tottenham Marshes, and the primrose is abundant at Harefield and Hadley, although it has been eradicated in the neighbourhood of London. It is interesting to note that Primrose Hill is said to have derived its name from the former abundance of *Primula vulgaris* in this locality.

The following extract from Richard Jefferies' *Nature near London* has reference to the flora along the Teddington Reach of the Thames. "Beneath the towing-path, at the roots of the willow bushes...the water docks lift their thick stems and giant leaves. Bunches of rough-leaved comfrey grow down to the water's edge—indeed, the coarse stems sometimes bear signs of having been partially under water when a freshet followed a storm. The flowers are not so perfectly bell-shaped as those of some plants, but are rather tubular. They appear in April though then green, and may be found all the summer months. Water-betony, or persicaria, lifts its pink spikes everywhere, tiny florets close together round the stem at the top; the leaves are willow-shaped, and there is scarcely a hollow or break in the bank where the earth has fallen which is not clothed with them. A mile or two up the river the tansy is plentiful, bearing golden buttons, which like every fragment of the feathery foliage, if pressed in the fingers, impart to them a peculiar scent. There, too, the yellow loose-strife pushes up its tall slender stalks to the top of the low willow bushes, that the bright yellow flower may emerge from the shadow....The river itself, the broad stream, ample and full, exhibits all its glory in this reach."

The ferns are naturally poorly represented in Middlesex, and have become of late years very scarce in the vicinity of London. The lady-fern and hart's-tongue are very rare, but may be found in the woods about Harrow, Highgate, and Harefield. The Osmunda, or flowering fern *O. regalis*, is found very rarely in bogs on heaths, and will soon be all gone; but the common fern is met with on heaths and open uncultivated places.

There is much that might be said of the trees of Middlesex, the oak, hawthorn, birch, beech, and chestnut being the better known. The district known as Enfield

Chase was, in Evelyn's days, an area of four thousand acres of forest, with some 630,000 trees, principally oak, ash, and beech. Now there are only a few patches of rough woodland to remind us of the sylvan beauties of that fine forest tract. The hornbeam, which is one of the most distinctive of Essex trees, once grew freely in Middlesex, and the labyrinth in Hampton Court grounds is formed of hedges of hornbeam. There are some fine oaks in Bushey Park, and two magnificent ones in Twickenham Park. Loudon records that the Chandos oak, which grew in the grounds of Michendon House, Southgate, was remarkable for its immense head, covering a space of ground 118 feet in diameter. The hawthorn is very common in all districts, being generally planted as a quickset hedge. There are, however, some old thorns in Bushey Park and on Hounslow Heath. Bushey Park is chiefly noted for its magnificent avenue of horse-chestnut trees, more than a mile long and unequalled in England. The trees were planted by William III, who was a great gardener, and are interspersed with flowering limes, which together form delightful avenues. The yew tree is found in hedges, and in some of the churchyards, especially those of Hanworth, Edmonton, Bedfont, and Harlington. At the latter place, the famous yew was once a masterpiece of the topiary art, and the clipping of the tree was made the occasion of a village festival. Now, at about 18 inches from the ground, it is 19 feet round its gnarled and knotted trunk. In writing of the trees of Middlesex, mention must be made of the Hampton Court Vine, which was planted in 1768. It

Chestnut Avenue, Bushey Park

is one of the largest in the country, and has produced as many as 2200 bunches in one season.

The wild animals of Middlesex are similar to those that exist in most English counties. More than twenty years ago Richard Jefferies wrote: "The list of really wild animals now existing in the Home Counties is so very, very short, that the extermination of one of them seems a serious loss. Every effort is made to exterminate the otter....Londoners, I think, scarcely realize the fact that the otter is one of the last links between the wild past of ancient England and the present days of high civilization. The beaver is gone, but the otter remains, and comes so near the mighty city as just the other side of the well-known (Teddington) Lock. The porpoise, and even the seal, it is said, ventures to Westminster sometimes, the otter to Kingston."

There are still several hundred head of fallow deer in Bushey Park. They are very tame and find sanctuary in the enclosed stretches of woodland and undergrowth. Enfield Chase, to which reference has already been made, was once the home of the deer, which Macaulay says "wandered there by thousands." Evelyn, writing on June 2, 1676, records that it was "stored with not less than 3000 deer."

Mr Dixon, who has recently written on the bird-life of this district, thinks that a lifetime among the birds might be profitably spent well within the fifteen-mile radius of London. Not only do the residents and regular migrants offer scope for prolonged study, but the casual visitors are a source of great interest in themselves.

There is no doubt that the migration of birds over Middlesex is on a much larger scale than is generally supposed, for this county lies in the direct path of many of these journeying birds on their way northwards from the Sussex coast. The parks and open spaces around Highgate, Edgware, Stanmore, Pinner, and Harrow are attractive haunts for birds, while the large reservoirs of Kingsbury and Elstree, and the rivers Thames, Colne, Brent, and Lea have a special charm for various species of waders and wild-fowl.

About 30 years ago it was estimated that there were 225 species of birds in Middlesex, of which 60 were classed as resident, 68 as migratory, and 97 as rare and accidental visitors. It is probable that the number of species is now less, although the variety of bird-life in Middlesex is still very great.

The peregrine falcon has been killed in Finchley, Highgate, and Harrow, and the kestrel may be seen at Osterley Park, Ealing, and Finchley. The sparrow-hawk breeds at many places within a few miles of London, and the barn owl haunts Harlesden and Kensal Rise. The blackbird is found in exceptional numbers about the gardens and orchards of Acton, Ealing, and Gunnersbury, while the song-thrush frequents almost every open space covered with trees and undergrowth. The nightingale is still heard in Osterley Park and West Middlesex, but there is no doubt that its numbers have decreased in recent years. The numerous Nightingale Lanes and Nightingale Roads in urban Middlesex suggest that nightingales once abounded but are no longer to be found. The chaffinch, the linnet, and the starling are next to the sparrow the most familiar birds of Middlesex. The raven is a rare visitor, although one nested in recent times at Enfield. The carrion crow still lives in the county, the rook is common in parts, while the jackdaw rears its young near some of the busy thoroughfares, one of its favourite resorts being around Harrow. The magpie and the jay have decreased in numbers, but breed at Osterley, Hendon, Ruislip, and elsewhere. The kingfisher is occasionally seen on the Thames west of Battersea, and frequents the Brent, the Colne, and the Lea. It is a local species, and its breeding places are very few, for this bird is now nowhere numerous. The ringdove is common and widely dispersed over the rural and outlying districts, especially in the west of the county. The lapwing is a fairly well-known visitor at Mill Hill, Stanmore, Pinner, Harrow, and Hendon, but the woodcock, once very common, is now rarely found on Wormwood Scrubs and about Harrow and Kingsbury. In passing, we may refer to the fact that there are at least three Woodcock Hills in the county. The water-hen or moorhen frequents spots on the Thames, the Brent, and the Lea, and is equally familiar at Pinner and Uxbridge; and the coot, somewhat rarer, is found at Osterley, Ruislip, and Wembley. It is of interest to note that the last heronry in the county was in Osterley Park, which has been so often mentioned in this chapter.

Some reference may here be made to the swan, which is a dweller on some of our ornamental waters, and numbers live in a semi-wild state on the Thames. This bird is of great historical interest to Londoners, and enters largely into the royal and civic life of the metropolis. Swans, according to the law of England, are royal birds, and when found on the sea and navigable rivers, are presumed to belong to the Crown. These royal birds on the Thames have five "nicks" or marks on them, and the king's swanherd was once an important person. Besides the swans belonging to the Crown, the Vintners' and Dyers' Companies are also permitted to keep a certain number. The Vintners' mark is two nicks, while each of the Dyers' birds has one nick. The "upping" trip now occupies four days, and the swans are marked from Southwark Bridge to Henley.

9. Climate and Rainfall.

The climate of a country or district is, briefly, the average weather of that country or district, and it depends upon various factors, all mutually interacting, upon the latitude, the temperature, the direction and strength of the winds, the rainfall, the character of the soil, and the proximity of the district to the sea.

The differences in the climates of the world depend mainly upon latitude, but a scarcely less important factor is proximity to the sea. Along any great climatic zone there will be found variations in proportion to this proximity, the extremes being "continental" climates in the centres of continents far from the oceans, and "insular" climates in small tracts surrounded by sea.

Continental climates show great differences in seasonal temperatures, the winters tending to be unusually cold and the summers unusually warm, while the climate of insular tracts is characterised by equableness and also by greater dampness. Great Britain possesses, by reason of its position, a temperate insular climate, but its average annual temperature is much higher than could be expected from its latitude. The prevalent south-westerly winds cause a drift of the surface-waters of the Atlantic towards our shores, and this warm water current, which we know as the Gulf Stream, is the chief cause of the mildness of our winters.

Most of our weather comes to us from the Atlantic. It would be impossible here within the limits of a short chapter to discuss fully the causes which affect or control weather changes. It must suffice to say that the conditions are in the main either cyclonic or anticyclonic, which terms may be best explained, perhaps, by comparing the air currents to a stream of water. In a stream a chain of eddies may often be seen fringing the more steadily moving central water. Regarding the general north-easterly-moving air from the Atlantic as such a stream, a chain of eddies may be developed in a belt parallel with its general direction. This belt of eddies, or cyclones, as they are termed, tends to shift its position, sometimes passing over our islands, sometimes to the north or south of them, and it is to this shifting that most of our weather changes are due. Cyclonic conditions are associated with a greater or less amount of atmospheric disturbance; anticyclonic with calms.

The prevalent Atlantic winds largely affect our island in another way, namely in its rainfall. The air, heavily laden with moisture from its passage over the ocean, meets with elevated land-tracts directly it reaches our shores—the moorland of Devon and Cornwall, the Welsh mountains, or the fells of Cumberland and Westmorland—and blowing up the rising land-surface, parts with this moisture as rain. To how great an extent this occurs is best seen by reference to the accompanying map of the annual rainfall of England, where it will at once be noticed that the heaviest fall is in the west, and that it decreases with remarkable regularity until the least fall is reached on our eastern shores.

The above causes, then, are those mainly concerned in influencing the weather, but there are other and more local factors which often affect greatly the climate of a place, such, for example, as configuration, position, and soil. The shelter of a range of hills, a southern aspect, a sandy soil, will thus produce conditions which may differ greatly from those of a place—perhaps at no great distance—situated on a wind-swept northern slope with a cold clay soil. The character of the climate of a country or district influences, as everyone knows, both the cultivation of the soil and the products which it yields, and thus indirectly as well as directly exercises a profound effect upon Man.

In considering the climate of Middlesex we must bear in mind that it is not a maritime county like Essex, and so has not the modifying influence of the sea. It will be well also to remember that, in point of size, Middlesex is

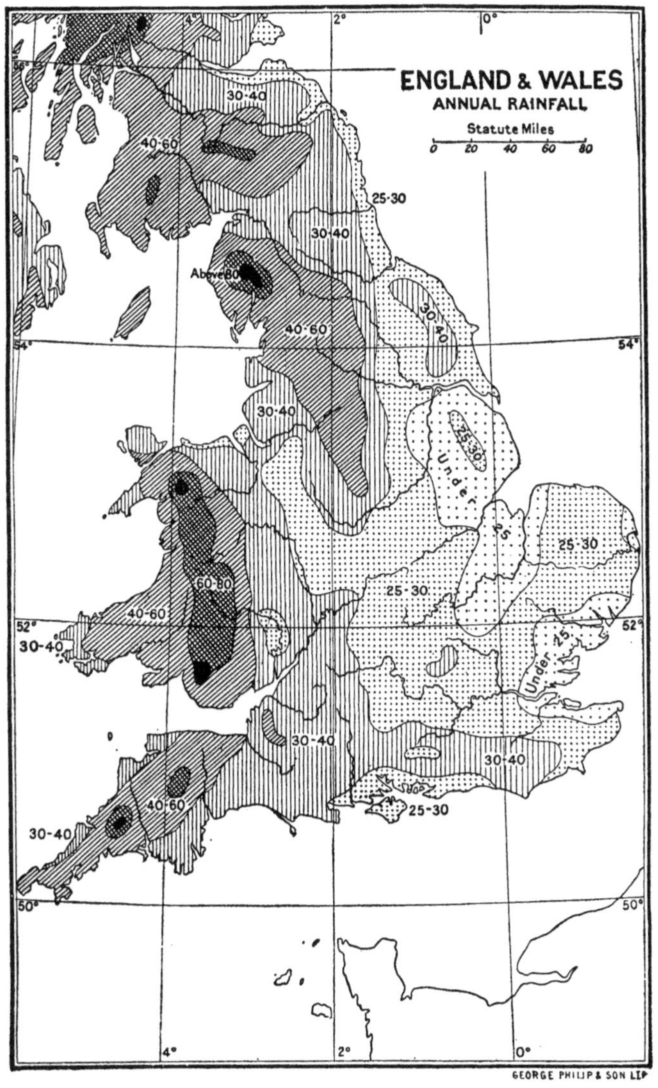

(The figures give the approximate annual rainfall in inches)

one of the smallest of our English counties, and so we must not expect to find the variations in its climate so noticeable as those in Kent or Essex.

It is of the greatest importance to have accurate information as to the prevailing winds, the temperature, and the rainfall of a district, for the climate of a county has considerable influence on its productions, and its trades and industries. Our knowledge of the weather is much more definite than it was formerly, and every day our newspapers contain a great deal of information on this subject. In London there is the Meteorological Society which collects information from all parts of the British Isles relating to the temperature of the air, the hours of sunshine, the rainfall, and the direction of the winds. The British Isles have been divided for these purposes into twelve districts, and day by day the newspapers publish the forecasts issued by the Meteorological Society of the probable weather for the twenty-four hours next ensuing ending midnight. Thus for October 27, 1910, the following was the forecast for Middlesex: "Easterly breezes, fresh at times in places; fair or fine to dull; mild." Warnings are also issued when necessary so that certain districts may be prepared for the rough weather that may be expected. Besides this information, some of the newspapers print maps and charts to convey the weather intelligence in a more graphic manner.

The prevailing winds of Middlesex, like those of the British Isles generally, are south-westerly. For a short period of the year, Middlesex suffers from the east wind, and during its prevalence in March and April, the air is dry, and catarrhal complaints are common. For two or three months in each year the county is liable to fogs, which are most dense and disagreeable in the districts near London, especially in the valleys of the Thames and the Lea.

The warmest month in Middlesex is generally July, when an average temperature of 60° prevails, and January, the coldest month, has an average temperature of 37·8°. The mean temperature of England was 47·7° in 1909, and that of Middlesex 48·1°. The northern and north-western parts of the county are generally colder and more bracing than those to the south. This difference mainly depends on the lower elevation of the southern as compared with the northern districts. Such places as Harrow, Highgate, and Harefield are noted for the salubrity of their climate, which is owing not only to their elevation but also to the nature of the substratum. During the year 1909, there were 1656 hours of bright sunshine in Middlesex, which compares favourably with 1565 hours for the whole of England.

Now let us look at the rainfall statistics for 1909, and for purposes of comparison it may be mentioned that the rainy days for England were 192 in 1909, and for the same year the rainfall was 30·31 inches. Middlesex is far below the average for the country, and this is largely due to the fact that the rainfall of England and Wales generally decreases from west to east. Thus we find that the highest rainfall in 1909 was at the Stye, in Cumberland, where no less than 176·30 inches were measured, and the least was at Southery in Norfolk, where the record was 21·37 inches. Now let us compare these two extremes with those of Middlesex. The highest rainfall registered in this county in 1909 was 28·76 inches at Rammey Marsh, and the lowest of 22·4 inches was at Sipson. Of course these results vary from year to year, but as they have been collected for a number of years, the average may easily be ascertained for any place that has a station for registering the daily rainfall. The rainfall of Middlesex for the year 1909 was estimated at 24·95 inches, and during the same period there were 179 days with rain. It is worth noting that February was the driest month, and October the wettest month throughout Middlesex in 1909, which was also the case in the whole of England.

We may summarise the climate of Middlesex by remarking that while this county is too small to have its own regional climate, yet it is sufficiently large to present many local shades of climate. In the first place we note that Middlesex belongs to the east coast district, and though it is relatively distant from the North Sea, the county of Essex affords little protection from the east winds which sweep over it in the spring. We also note that Middlesex is open on the south-west and is influenced by the south-westerly winds from the Atlantic. Thus the climate of Middlesex is a mixed and tempered product of two influences, and its changes from relaxing or "muggy" to dry and bracing may be readily understood. The openness of the valley of the Thames, and the absence of great neighbouring heights, also keep it from excessive cloud and rainfall. The chief characteristic of its climate is perhaps the small range of temperature. Thus over a period of 80 years the winter temperature was 37·6° and the summer temperature was 60°, showing a range of only 22·4°. A night on which the thermometer remains for some hours below zero is of the rarest occurrence, the lowest recorded being – 6·5° on Dec. 24, 1796.

10. People—Race. Dialect. Settlements. Population.

The earliest people of whom we have any traces in Middlesex are the people of the Palaeolithic, or Old Stone Age. Palaeolithic man belonged to a type with which the Eskimos and Bushmen have been compared as modern representatives. He did not cultivate the land, but lived on the food which nature supplied. We know little of him, except from the rough unpolished stone implements that have been found in many parts of Middlesex, and of course in other districts as well. The Palaeolithic were followed by the Neolithic, or New Stone Age people, and these early immigrants are considered to have come from the area around the western portion of the Mediterranean. They are often called Iberians, and knew how to grind and polish flint and other hard stones. Later came the Goidels and Brythons, some of whom may have settled in the Middlesex area. They were acquainted with bronze and iron, and many of them cleared the woodlands and became herdsmen and tillers of the soil. There are few written records about these early people, but when Julius Caesar invaded our land in 55 B.C. he found the Britons belonging to various races, using different languages, and in various stages of civilization. The people in Middlesex were the Trinobantes, a branch of the Celts, and from the accounts we have there is no doubt that they were skilful in war and were under the leadership of brave chiefs.

It seems probable that Middlesex formed but a small portion of the country peopled by the Trinobantes, whose territory included Essex and Hertfordshire. The only Celtic place-names surviving in the Middlesex of to-day are London and Thames, and there is no doubt that its conquest by the Romans was complete, and that the natives were Romanised in many ways. The Roman invasion introduced a small but powerful element of very diverse origin, for there were dark-haired Iberians from Italy, Spain, and Southern Gaul, and red-headed Celts and flaxen-haired Belgi in the Roman armies. When the Romans left Britain at the beginning of the fifth century, Middlesex fell an easy prey to the Teutons and Scandinavians, who were flaxen or red-haired and blue or grey-eyed northern Aryans. The Angles, Frisians, and Saxons came over in bands, and settled at various times in separate localities. The area that formerly belonged to the Trinobantes seems to have been chosen by the East Saxons, and as time went on a section of this people settled between the West Saxons on one side and the East Saxons on the other side. They called themselves Middle Saxons and gave their name to the county. The speech of these invaders became general, and so thorough was the Saxon conquest of Middlesex, that nearly all the places received new names, which are retained to this day. Nothing shows more clearly the complete character of this conquest than the place-names. Fields, fords, hams, and greens are numerous; and such places as Highgate, Southgate, Acton, Ashford, and Finchley tell their own story in their names. The population must have been very small, even down to the time of the Norman Conquest, and the inhabitants chiefly settled along the line of the old Roman roads, on the banks of the Thames, and in some forest clearings.

A further great change was effected in 1066, when the descendants of the Northmen, or Vikings, who had settled in Normandy, conquered the land of the English. Many of these Norman invaders were largely of French extraction and of the dark-haired type; but afterwards numbers of Gascons, Poitevins, and other Frenchmen entered the country and held positions of influence and importance. William the Conqueror imposed his will on the people, and as far as Middlesex is concerned we can see how thorough was his conquest by a reference to the Domesday Book. There we find that the county was in the hands of 30 tenants-in-chief, holding their estates direct from the Crown, the most important being Geoffrey de Mandeville.

The Norman invasion marks an epoch in our history. Since the eleventh century there has been no hostile invasion of our land, but there have been periods when foreigners have come over and settled in Middlesex. The Jewish type has exerted considerable influence, especially in the suburbs around London; and during the latter portion of the nineteenth century many thousands of aliens came from Germany, Russia, and elsewhere, and settled largely at Tottenham, Hornsey, and Willesden.

From the foregoing remarks it will be gathered that the people of Middlesex are mainly of Teutonic stock

Foreign Quarter, Tottenham

and of English speech. After the Norman Conquest the English language had at least six dialects, and Middlesex comes within the domain of the south-eastern dialect. During the course of all the centuries this dialect has been so modified and altered by the prevailing speech of London that we can no longer speak of the Middlesex dialect. Nothing has done more to spoil the speech of

Middlesex proper than the infusion of the "Cockney" element, which has worked sad havoc in the parishes around London.

Having considered the facts relating to Middlesex in the past, we may turn our attention to the people of this county as we find them to-day. It will be well to remember that we are now dealing with the county of Middlesex when it is considerably smaller than it was before the County of London was formed in 1888.

The population of the Administrative County in 1911 was 1,126,465, which shows an increase of 333,989 or about 42 per cent. in the last ten years. The density of population is very striking, for whereas the average population of a square mile in England and Wales is 618, in Middlesex it is 4855.

This great increase in the population of Middlesex has been mainly in the districts near London, such as Tottenham, Willesden, Ealing, Hornsey, Harrow, Edmonton, and Enfield. These districts are almost entirely residential, and are connected by railways and the electric tramway system with London. If we look at the figures relating to Willesden we find that its population of 751 in 1801 grew to 114,811 in 1901 and to 154,214 in 1911, and this increase is typical of the other districts which have become thoroughly urbanised. The urban population of Middlesex in 1911 was 1,078,334 against the rural population of 48,131. The bulk of these people lived in houses or tenements, of which 103,318 contained five or more rooms, and 62,656 had less than five rooms.

The census figures have many interesting facts. In

1911, the females in the county exceeded the males by 75,600; and while in 1901 there were 30,973 people over 65 years of age, there were only two centenarians. We also find in 1901 that 266,563 were born within the county, 241,384 in London, 22,962 in other parts of the British Isles, 6241 in British Colonies, and 9700 in foreign countries, chiefly Germany, France, Italy, United States, and Russia. A reference to the diagrams at the end of this book will show that the population of Middlesex advanced from 171,024 in 1861 to 792,476 in 1901, and to 1,126,465 in 1911.

11. Agriculture—Main Cultivations, Woodlands, Stock.

Although Middlesex is one of our smallest English counties and has an enormous urban population, it is of some importance from an agricultural point of view. The population is chiefly concentrated on the borders of the metropolis; and the portions of the county remote from London, especially those in the north-west, south-west, and north-east, have a comparatively scanty population, of whom a large number are employed on the land.

It will be well at the outset to refer to the diagrams dealing with agriculture at the end of this volume, and form in our mind's eye a picture of the relative areas growing the various crops. The chief fact we must first grasp is that 88,834 acres of Middlesex are under crops and grass, and that this acreage is about three-fifths of the area of the whole county. We must also remember that Middlesex is no longer a great wheat-producing county as it was in the early part of the nineteenth century, when no less than 10,000 acres were growing wheat. The latter portion of that century saw a great change in the agricultural condition of Middlesex, for it was then

Market Garden, Isleworth
(*Showing vegetables being forced under clochés*)

found more profitable to lay the land down in grass than to grow corn.

We will now proceed to consider the chief crops grown in Middlesex, and this we can do with the assistance of the report for 1911 of the Board of Agriculture. In that year it was estimated that the area of this county, excluding water, was 147,007 acres, and, as we have already said, 88,834 acres were under crops and grass. The corn crops were wheat, barley, oats, rye, beans, and peas, which were grown on 6417 acres, or only about one twenty-second part of the whole county. Wheat and oats were the most important of the corn crops, and together account for 4554 acres of the small area under corn. It will thus be seen that the remaining corn crops are of little importance. The contrast with the adjoining county of Essex is most striking, for there no less than one-fourth of its area is devoted to the growth of corn against the 6417 acres in Middlesex.

The green crops, comprising among others, potatoes, turnips, mangold, cabbage, and vetches or tares, are grown on 11,077 acres, while clover, sainfoin, and grasses under rotation claim 2528 acres. The largest part of the agricultural area of Middlesex is under permanent grass, which accounts for 64,838 acres, or nearly one-half of the county. Small fruit is a rapidly-increasing product, and has reached an area of 3974 acres and other fruit grown in orchards accounts for 5389 acres. The extent of woodlands has much diminished in the county, and now the acreage of coppices, plantations, and other woods is 3968, while 1342 acres are under bare fallow.

The cultivation of fruit and vegetables for the London market has always been of great importance, and although the urbanisation of so much of Middlesex is extending, fruit-farms are numerous in the neighbourhood of Twickenham, Brentford, and Isleworth. West Middlesex is almost wholly given over to market gardening, the produce, comprising fruit, root-crops, and vegetables, being conveyed by road to London. No sight is more familiar to Londoners than the great wagons, loaded with empty baskets, returning through the chief thoroughfares to their destination in rural Middlesex after conveying their produce to Covent Garden and other markets. Here again we see the dependence of Middlesex on London for its support from an agricultural point of view.

Glasshouses, Isleworth

We can get a good idea of the fruit and vegetable culture by going on the tram beyond Brentford. On the Isleworth side we pass through wide orchards, and by acres of forcing houses and nursery grounds. Around Brentford are many market gardens and wide stretches of orchards, while hundreds of acres are devoted to the raising of rhubarb and various vegetable crops. Here it is no uncommon thing to see women engaged in gathering and bundling the crops for the London markets. A hundred years ago it is recorded that the fruit harvest from Isleworth orchards was carried by women to Covent Garden in weighty loads—"these laborious females sustain their burden on their head." Since then things have changed, but even now women play their part on the soil of Middlesex with profit to themselves.

Middlesex, too, has of late years developed its dairy-farming, and tracts once famous for the best wheat are now devoted to pasturage. In the time of Queen Elizabeth, Middlesex wheat was of great repute, and we read that the royal manchet or best white bread was made of flour from grain grown at Heston. Now, that district is almost given up to pasturage.

In concluding our review of the Middlesex Agricultural Report we may notice the domestic animals. These are classified as horses, cows and other cattle, sheep, and pigs. Of these, in 1911, sheep were most numerous and numbered 19,001. Cows and other cattle numbered 17,337, pigs 18,059 and horses 6718. Cows are reared in large numbers to supply milk for London and suburban consumption, and while the home producer retains his present monopoly of the supply of fresh milk to the community, there is no doubt that dairy-farming will be very profitable for the agriculturists of the country. Shorthorns and Channel Islands cattle are kept by all the leading landowners for dairy purposes. Excellent Down sheep are to be seen on the farms and the best breeds of pigs are reared.

12. Industries and Manufactures.

It will be gathered from the previous chapters that Middlesex is largely a residential county for the people who work in London, where nearly all the manufactures and commerce are carried on. The purely local industries, though not very important, are increasing, and a good many people are employed in agriculture, to which reference was made in the preceding chapter.

Although there is no great staple industry, there are numerous factories to supply some of the wants of London. There are soap-works at Chiswick and Brentford, but especially at Isleworth. There are asbestos works at Harefield, and electric-light apparatus is manufactured at Ponder's End. Linoleum is made in large quantities at Edmonton and Staines, while Willesden has attained considerable notoriety for its manufacture of canvas. Brentford, Isleworth, Chiswick, and Staines are some of the chief brewery centres, and mineral waters are made at Tottenham and Uxbridge. There are flour-mills at Isleworth, and at some places on the Colne and the Brent. Powder, snuff, and paper-mills have long flourished on the banks of the Crane, and gunpowder is extensively made in the well-known mills near Hounslow. Ammunition is made at Hanworth, Edmonton, and Hendon, and chemicals are produced at Southall and Greenford. The chemical works at Greenford are of great importance, for it was there that Sir William Perkin carried out his discovery of producing aniline dyes from coal-tar, a discovery which has revolutionised the art of dyeing woollen and cotton stuffs, as well as silk. India-rubber is manufactured at Yiewsley, Tottenham, and Wembley, and there are varnish factories at Edmonton. Iron-foundries of more or less importance still exist at Uxbridge, Brentford, Twickenham, Isleworth, and Staines, and there are engineering and other works at Hayes. West Drayton is noted for the manufacture of granolithic stone, and swords are made at Chiswick.

Besides this miscellaneous collection of industries, there are a few others that deserve our attention. There are motor-car works at Acton, and boat-building and hiring afford considerable occupation at such river-side towns as Chiswick, Isleworth, Twickenham, and Teddington. In this connection Messrs Thornycroft's famous works at Chiswick must be specially mentioned. Besides turning out steam-launches and torpedo-boats in large numbers for the English and foreign governments, this firm has built in recent years some powerful gun-boats for the Royal Navy.

Enfield has the Government small-arms factory, and gave its name to the once famous Enfield rifle. The Enfield factory is the establishment through which most of the small-arms of every description have been supplied since the Crimean War to the army. Its long ranges of buildings and tall chimneys are conspicuous objects, and give one some idea of the importance of the work that is carried on so successfully. It may be mentioned that the Enfield rifle was the weapon of the British army from 1853 to 1865, and was then superseded by the adoption

Thornycroft's Works, Chiswick

of breech-loading arms, such as the Snider, the Martini-Henry, and the present magazine rifle. Besides the extensive factories at Enfield there are about 40 acres of land used for testing purposes. As many as 100,000 rifles can be manufactured annually, besides a large number of pistols and machine guns; and under great

Entrance to Enfield Small-Arms Factory

pressure this number, large as it is, could be greatly exceeded. The interior of the factory consists of large work-rooms, while several buildings are occupied as stores of stocks, barrels, etc., which are kept constantly in readiness for use as required.

Middlesex has very extensive brick-yards, more especially in the Thames valley brick-earth and in the London

Clay. In many places the level of the land has been considerably lowered by excavations for brick-earth, which has cleared off large tracts, exposing the gravel beneath, and this process is carried on at the present day at West Drayton, Hayes, and Southall. Near Hayes the bricks are made of a mixture of brick-earth, ashes, and chalk. They are first sun-dried, and then stacked in kilns and burnt. The brick-earth having been removed, the gravel is then sold at so much an acre, and worked to within a foot of the water-level. Although the main works lie west of Southall and Heston, brick-yards are also worked at Edmonton and Enfield in the north, and at Shepherd's Bush and Acton in the west.

13. History of Middlesex.

At the dawn of British history Middlesex was an almost uninhabitable waste. The northern and eastern parts of the county were covered with a dense forest, while in the south the tidal waters of the rivers spread themselves over a wide area of marshland. Here and there in these marshlands slight mounds and banks of drift gravel formed the earliest settlements of the Britons who lived here when Julius Caesar invaded our land. The Trinobantes were the British tribe who inhabited an area which we now call Essex, Hertfordshire, and Middlesex. They were under the leadership of a brave chief, Cassivellaunus, who led a great force of Britons to oppose the advance of the Roman emperor. In the early skirmishes the heavily armed Roman soldiers suffered severely from the dashing onslaught and rapid retreat of the British chariots and cavalry. It would appear that Cassivellaunus gathered his forces on the north bank of the Thames, and Caesar had to cross the river at one of the few fords then in use. Most historians are agreed that the place selected was Cowey Stakes, a little to the east of Halliford. We cannot do better than give an

Halliford

account of the conflict in Caesar's own words. He says in the *De Bello Gallico* that the river was passable on foot only at one place, and that with difficulty. When he came there, he observed that there were large bodies of the enemy drawn up on the opposite bank. The bank also was defended with sharpened stakes fixed in front, and stakes of the like kind were fixed below under water, and concealed by the river. Having learnt thus much from the prisoners and deserters, Caesar sent forward the cavalry and immediately ordered the legions to follow them; but the soldiers went at such a pace and with such an impetus, though they had only the head above water, that the enemy could not resist the onset of the legions and the cavalry, but deserted the bank and took to flight. We need not go further into details, except to mention that many of the Britons surrendered to Caesar, and when he followed after the enemy to Verulamium, the present St Albans, he gained a complete victory. So thorough was the defeat of the Trinobantes, that Cassivellaunus was forced to offer his submission, which Caesar readily accepted.

Cowey Stake

It is not necessary to proceed further with the conflicts between the Britons and the Romans, more especially as our knowledge of this period is not very definite. There is no doubt that the Romans finally subjugated this part of Britain, and by the fourth century the district we now call Middlesex was included in the Roman province of Flavia Caesariensis. Augusta, as London was called by the Romans, became one of the chief towns in our land, and after its foundation several main roads were made, branching out from it to the east, north, north-west, and west, and so bringing Middlesex to a great extent under cultivation and within reach of Roman civilisation. The work of the Romans in Middlesex and references to their remains will fall into a later chapter.

When the Romans withdrew from Britain in 410 A.D. this county fell an easy prey to the Saxons, Angles, and Jutes. The counties we now call Essex, Hertfordshire, and Middlesex were attacked by the Saxons, and formed into the kingdom of the East Saxons about A.D. 492. Erkenwine was the first of fifteen kings of Essex, and he began to reign about A.D. 527, having London for his capital. Middlesex is not mentioned in our history till the eighth century, and in A.D. 704 we find it only as *provincia. Middelseaxan*, as it was then called, is not once named by Bede, and there is no evidence that it formed a separate kingdom till a later period. We find Ethelbald King of the Mercians making a grant of land in the province *Midelsexorum*, and in 767 it occurs in a charter as *Middil Saexum*. These early references are of the greatest value, as we infer that the name Middlesex signifies the country of the Middle Saxons, and was the territory which fell to one of the kings when Essex, which had been conterminous with Wessex, was divided. Eventually Middlesex was annexed by Mercia, and though the date of this event is unknown, it would seem to have taken place before the end of the eighth century. London was a place of meeting for the Mercian council from 748, and Brentford from 780.

When Essex passed under the West Saxon rule in 825, there is reason for believing that Middlesex remained Mercian, for in a charter of 831 the King of Mercia made a grant of land at Botwell "*in provincia Middelsaxanorum*" to the Archbishop of Canterbury.

We are almost forced to the conclusion that Middlesex was never an independent kingdom, and as it is only once mentioned in the *Old English Chronicle* it must have played an insignificant part in those early years of our history. In the ninth century it was overrun by the Danes who were eventually defeated by Alfred, who made terms with them at Wedmore in 878. By this treaty the river Lea became the boundary between the Saxons and the Danes, and thus Middlesex was formally separated from Essex, and became an independent county. Again at the beginning of the eleventh century the Danes crossed the Thames at Staines, and ravaged the county; and in 1016 Canute was defeated at Brentford, then an important town, by Edmund Ironside.

During the period of our Saxon history it is evident that the Saxons settled over large parts of Middlesex and made clearings in the great forest. By the time of Edward the Confessor a large proportion of the present towns and villages were in existence, but London had so grown in importance as to overshadow the shire of which it was once nominally the capital. "The shire was let to the men of London and their heirs to be held in favour of the King and his heirs; and the subject shire was to submit to the authority of the sheriffs chosen by the ruling city." Here we see stated in formal terms the subjection of the shire to the capital, which condition lasted till the present county of London was formed in 1888.

When the Norman Conquest was completed by William, Middlesex was divided into six Hundreds; and from the Domesday Book of 1086 we learn something about the chief men in the county. Middlesex then had 24 tenants-in-chief holding land direct from the King, Geoffrey de Mandeville being the most important. The greater part of the county, however, belonged to the Church, the chief owners being the Archbishop of Canterbury, the Abbot of Westminster, the Abbess of Barking, and the Bishop of London.

From the time of the Norman Conquest the great forest of Middlesex is often mentioned by our early historians, and even as late as the time of Elizabeth, large tracts of it existed close to London. In Norman times, Fitzstephen mentions the existence of an immense forest, having densely wooded thickets and coverts, the haunt of red deer, fallow deer, boars, and wild bulls. In 1218 the forest was disafforested, and wealthy London citizens purchased some of the land and built on it. Matthew Paris describes the woods contiguous to Watling Street, between London and St Albans, as almost impassable, and so much infested by outlaws and wild beasts that pilgrims to the shrine of St Alban were exposed to great dangers.

From the twelfth century downwards, the history of Middlesex is little else than the history of the metropolis, and as that is dealt with in the volumes on *London* in this series, we need only refer to a few of the more noteworthy events which have taken place in Middlesex outside the limits of its former capital.

In 1217, at Hounslow, a conference was held between four peers and twenty knights, on the part of Louis the Dauphin, and the same number of nobles and knights on the part of the young King, Henry III. Later in the same reign, Simon de Montfort and the barons encamped at Isleworth; and four years later, in 1267, the people of London were marshalled on Hounslow Heath by the Earl of Gloucester to give battle to Henry III.

The Duke of Gloucester with other nobles met Richard II at Hornsey in 1386, and compelled the young king to dismiss his favourite, Robert de Vere. During the insurrection of Jack Cade in 1450 the Essex insurgents met at Mile End; while in 1461 the Kentish insurgents did some damage at Highgate.

One of the most famous battles in our history was fought in 1471 at Barnet, on a tract of ground which is included in Middlesex. The Earl of Warwick, who had changed his allegiance from the Yorkists to the Lancastrians, encamped his troops on Gladsmoor Heath to the north-west of Monken Hadley Church. The kingmaker's opponent, Edward of York, also camped on the heath, and on Easter Day, April 14, the two armies gave battle. A fierce fight, stubbornly maintained by both sides, was turned by a fog to the advantage of Edward. The kingmaker seeing that all was lost, for most of his supporters were slain, withdrew towards Wrotham Wood. At the edge of that forest, with his back to a tree, he was surrounded by the enemy, and fell in his forty-fourth year. The dead trunk of an old elm remains as the traditional site of Warwick's death, and at the junction of the Hatfield and St Albans roads an obelisk stands to commemorate the decisive battle of Barnet.

When we come to the Tudor period, we find that Hampton Court, the favourite palace of Henry VIII, played a considerable part in the reign of that king. It

Monument at Hadley, near Barnet

(*In commemoration of the battle between Edward IV and the Earl of Warwick*)

was at Syon House, Isleworth, that Lady Jane Grey reluctantly accepted the crown, and was thence conducted with great pomp to London. In 1586, at Okington near Harrow, Anthony Babington and his fellow conspirators against Elizabeth were arrested. James I was met in 1603 at Stamford Hill on his way to London and was led with great ceremony of state to the Charterhouse.

The reign of this king is of interest to Middlesex, for on January 14, 1604, began the celebrated conference between the presbyterians and churchmen, before the King himself as moderator, in the privy chamber at Hampton Court. The conference lasted three days, and as a result a new translation of the Bible was ordered, and

Interior of the Treaty House, Uxbridge

some alterations were made in the Book of Common Prayer.

At the beginning of the Civil War, on November 12, 1642, the parliamentarians were defeated at Brentford by Charles I, and the eccentric John Lilburne, with four hundred men, was made prisoner. It was at Uxbridge, in January, 1645, that the fruitless negotiations between the King's and the parliamentary commissioners began, and were carried on for eighteen days. General Fairfax had his headquarters at Isleworth in 1647, and in the same year Charles I was kept in confinement at Hampton Court for nearly three months. Oliver Cromwell returning from his victory at Worcester, was met at Acton by the Lord Mayor and aldermen of London and many members of parliament, and thence accompanied to London in great triumph. When the Restoration was ushered in, General Monk assembled his army on Finchley Common; and in 1686 James II kept an army on Hounslow Heath to overawe London, but he broke up this camp in 1688.

In the Georgian period, Finchley Common was the camping-ground for the defence of London when the young Pretender was marching south on the capital. The political history of Middlesex in the eighteenth and nineteenth centuries was disturbed by the opposition to the election of Wilkes in 1768, 1769, and 1774, and by the riots in connection with the election of Sir Francis Burdett at a later period (1804–1806). During the nineteenth century Middlesex had little history of its own, and we may bring this narrative to an end by noting that the county was deprived of much of its area and a large portion of its population by the Act of 1888, by which the County of London was formed.

14. Antiquities—Prehistoric, Roman, Saxon.

The earliest evidence of the presence of man in Middlesex is not derived from written records, but from the antiquities, such as flint implements, that have been found in various parts of the county. Antiquaries have divided the earliest periods of our country's history into the Stone Age, the Bronze Age, and the Early Iron Age. These three periods cover a wide extent of time, but it is impossible to say how many thousands of years are included in each of them, for we are not certain when one age ended and the next began. It must not be thought that stone was discarded for many purposes in the Bronze Age, or that bronze was no longer in use in the Iron Age. On the contrary, the use of each material survived long into the succeeding period; but this classification of the prehistoric period of our country's history is convenient, as it shows the prevailing material that was in use and thus marks the development of man in various kinds of handiwork.

The Stone Age has been subdivided into the Palaeolithic or older section, in which the flint implements were formed only by chipping, and the Neolithic, or newer section, in which the implements were more carefully worked, and even polished. There are good reasons for the supposition that, in our country, an immense period of time separated the Palaeolithic from the Neolithic period, for geology comes to our aid and helps us to understand the length of time that was required to form the deposits in which the Neolithic implements were found.

As Middlesex mainly lies in the valleys of the Thames

Palaeolithic Flint Implement found in Gray's Inn Road

and the Lea, it is not at all remarkable that so many traces of early man are found within the borders of the county. The occurrence of Palaeolithic implements has been known for more than 200 years, and it is of interest to note that the first recorded discovery of a flint implement was made in London, when the capital of Middlesex, towards the end of the seventeenth century. This fine pear-shaped implement was found near the present Gray's Inn Road, and was at first described, wrongly of course, as a British weapon.

Palaeolithic implements have been found at Bush Hill Park, at Acton, West Drayton, and elsewhere. Almost all the implements have been fashioned out of flint, many from gravel-flints and others from flints taken out of the Chalk. Evidence of "Palaeolithic floors" on which the makers of flint implements had worked, were discovered in 1878, in the neighbourhood of Stoke Newington, and later in the Lea valley and at Acton. The late General Pitt-Rivers made some discoveries in the districts around Ealing and Acton, and the implements found were flakes of large size and bold workmanship. Sir John Evans, in his *Ancient Stone Implements*, gives more than twenty localities in Middlesex where Palaeolithic remains have been found. In the bed of the Thames, too, a good many specimens have been frequently brought up in the course of dredging operations. Many conjectures have been made as to Palaeolithic man himself, but although there is a wide diversity of views on this subject, we may say that he probably belonged to a type of civilization similar to that of the Eskimos, the native Australians, and the Bushmen of to-day. He did not cultivate the land, but lived on the food which nature supplied. With the exception of *Bos longifrons*, which has been found in the lowest terrace gravel at Twickenham, there has been no definite evidence of any domestic animals living at this period.

During the Neolithic Age, Great Britain was no longer joined to the Continent and was an island. The climate had become more temperate and rather moist, and such monsters as the mammoth had become extinct. Man had now learnt to train animals for domestic use; and he cultivated cereals for food, and various plants to provide materials for woven garments. He used the bow as his weapon, and he had also developed the art of making pottery. In the Neolithic Age, the implements were commonly hafted and made in a greater variety of forms; and by the addition of grinding and polishing, it was found possible to use other hard stones as well as flints. While the grinding and polishing of stones may be considered the special characteristic of this period, it must not be supposed that this was always the case. For instance, a large and important class of implements and weapons, such as knives, scrapers, and arrow-heads were but rarely ground or polished, while even axes of fine workmanship were sometimes finished by simply chipping them. Many celts of the Neolithic Age have been dredged from the Thames, and other implements and weapons have been found in various localities in the soil. A polished celt of stone was discovered at Northwood, and a perforated hammer of stone at Twickenham.

The enormous period known as the Stone Age, which can only be measured by geological time, left man still ignorant of many important arts of life. The introduction of metal marks one of the most important steps in human progress, and the period from the beginning of metallurgy down to the opening of recorded history is generally divided into two parts, named after the metals which successively occupied the most prominent place in human history. Thus we speak of an earlier, or Bronze Age, and a later, or Iron Age. These convenient terms have passed into general use, but they are not separated by a hard and fast division of time, nor do they exactly describe the work of the particular period.

The beginning of the Bronze Age in Great Britain is placed about 1800 B.C. and Sir John Evans has divided our Bronze Age into three periods:—(1) the period of the barrows, characterised by primitive forms hardly ever found apart from burials; (2) the period of the flanged celt and the tanged spear-head; and (3) the period of the bronze hoards, when swords and socketed celts and spear-heads are most conspicuous. The most important antiquities of the Bronze Age in Middlesex have been found at Hounslow. They comprise a flat, early form of celt, a sword in fragments, a socketed celt, and a palstave. There is no doubt that, at this early period, the people lived chiefly on the banks of the Thames, and this probably accounts for the fact that so many relics of the Bronze Age have been found in the bed of that river. Many of them are to be seen in the British Museum, where there is a particularly good example of a sword scabbard and another of a bronze buckler. Other parts of the county have also yielded relics of this age. Thus a socketed knife from Edmonton by the Lea, a spear-head from Hampton Court, a cinerary urn from Ashford, and a dagger from a barrow at Teddington may be mentioned. A visit to the British Museum will enable the student to realise what antiquities belonging to the Bronze Age have been found in Middlesex, and he will then be able to compare them and their workmanship with those obtained from other districts.

The Early Iron Age may be said to date approximately from 1000 B.C., and it has been said that the

Bronze Dagger, found at Teddington, 1854

antiquities of this period found in Great Britain present many features of interest and include several works of art that have never been surpassed in their particular sphere. As far as Middlesex is concerned the Early Iron antiquities have been found chiefly near the Thames or in the bed of the river. Perhaps the most noteworthy is a fine cruciform bronze mount from the Thames. It belonged most likely to a breast-plate, and has a depression in the centre, and ornamental discs on the side limbs. A series of bronze figures in the round were discovered at

Hounslow, and include three boars and two other non-descript animals, one with a loop for suspension. It is

Cruciform Bronze Mount, found in the Thames

probable that these miniature animals were crests of helmets.

The early camps and earth-works in Middlesex belonging to the pre-Roman period are not of much importance. An extensive fragment of an early camp may be seen near Enfield, while the defensive enclosures at Harmondsworth, Isleworth, and Twickenham have been almost obliterated. A great barrier known as Grims Dyke stretches from near Pinner Hall to the high ground of Harrow Weald. It has been suggested that this is a line of massive earth-work of vallum and fosse fashioned by the Britons to withstand the assaults from the south-east. Another view is that Grims Dyke is simply a tribal boundary of a somewhat later period.

The coming of the Romans to Britain marks the beginning of what may be called the historic period of our country. Outside the borders of the present County of London, the remains of the Romans in Middlesex have not been very extensive, and the only certain Roman station in the county was that of *Sulloniacae*, the present Stanmore. Some authorities think that Staines is on the site of *Ad Pontes*, and that there were Roman stations at Shepperton, Kingsbury, Bedfont, and elsewhere. We have already mentioned a ford of the Thames known as Cowey Stakes, where according to some of our antiquaries, Julius Caesar crossed the river during his second invasion of our land. In the days of the Romans, London was the chief town in their occupation, and Roman remains found in the City of London are very numerous. Reference to them is made in the volumes on *London* in this series; and in the Guildhall Museum there is a very fine collection of these antiquities. Outside London, finds of

Roman relics and coins have been made near East Bedfont, at Stanmore, and at Brockley Hill. The Roman roads made in Middlesex will be considered in the chapter on communications.

When we pass from the Roman period to the time of the Saxons, we have evidence that the English conquest of Middlesex was very thorough. Nearly all the names of places in the county are of Saxon origin and that fact alone speaks more eloquently than the finding of many relics beneath the surface of the soil. Acton, Ashford, Enfield, Greenford, Isleworth, Southgate, and Uxbridge are a few of the numerous places with Saxon names, and a glance down the parishes of this county will show how numerous are the "fields," "fords," "hams," and "greens." In Saxon times, the population must have been extremely small, and if we may judge by the size of the parishes, it is evident that the people were few and far between.

15. Architecture (*a*) Ecclesiastical—Churches and Religious Houses.

The churches of Middlesex were never on a grand scale, and a large proportion of them have been either wholly rebuilt in a modern style, or much altered and even mutilated in order to provide for the rapidly increasing population. There is no doubt that London completely overshadowed Middlesex from an ecclesiastical point of view, and so we find no great centre of church work and influence such as Canterbury or Rochester in Kent. Middlesex, however, has a number of interesting churches which date from the earliest Norman period and present numerous points of architectural interest. The churches at Harrow, Ruislip, Stanwell, Hayes, Harefield, and Harmondsworth are all worthy of consideration, either from their situation, their monuments, or their quaintness. With regard to the antiquity of the Middlesex churches, there is every reason to believe that they were all over the county at least 200 years before the Conquest; and although little remains of the original buildings, the present churches stand on the sites that have been hallowed through all the centuries of our history.

In the construction of its churches, Middlesex has suffered from the absence of building stone. As we have read in earlier chapters, Middlesex is on the London Clay with outcrops of chalk, and for many centuries the county was covered by forest. Hence we find that wood, brick, chalk, and flints are the materials generally used in the old churches. Clunch from Hertford and Surrey, and Ketton stone from Rutland are found in some of the churches, but stone was more freely used in the riverside churches, as it was brought up the river from Caen in France. Purbeck marble from Dorsetshire was brought here in large blocks for the fonts at Harrow, Harmondsworth, Ruislip, and elsewhere, and it was also used for decorative works such as shafts and arcading, as well as for monuments in the churches.

The churches are of all styles from what is known as Saxon to Perpendicular. It is often difficult to assign the exact date for the foundation of a church, as in many cases the early records are lost; and sometimes an old church of wood gave place to one built of more durable materials. Bede tells us that in his day there was not a stone church in all the land, but that the custom was to build them of wood. Hence there is no doubt that the early wooden churches were destroyed by fire or some other cause, and the present churches may stand on their site.

Architects and antiquaries are thus forced to assign the date of the building of a church by a careful study of its style of work. There seems a general opinion among recent authorities to class the churches at Kingsbury, Cowley, and East Bedfont as pre-Norman or Saxon buildings; and it is also thought probable that parts of the towers of Hendon and Hayes have work of this early period.

When we come to a later period we are on safer ground, and good work of the Norman period is to be found in the south doorways of Harlington and Harmondsworth, in the tower and west doorway of Harrow, and in the chancel arch at East Bedfont. Fragments of Norman work are also in evidence at Edmonton and Friern Barnet.

Towards the end of the twelfth century the round arches and heavy columns of Norman work began gradually to give place to the pointed arch and lighter style of the first period of Gothic architecture which is known as Early English, conspicuous for its long narrow windows, and leading in its turn by a transitional period into the highest development of Gothic—the Decorated period.

Norman Arch, East Bedfont Church

This, in England, prevailed throughout the greater part of the fourteenth century, and was particularly characterised by its window tracery. The Perpendicular, which, as its name implies, is remarkable for the perpendicular arrangement of the tracery, and also for the flattened arches and the square arrangement of the moulding over them, was the last of the Gothic styles. It developed gradually from the Decorated towards the end of the fourteenth century, and was in use till about the beginning of the sixteenth century.

There are examples of all these styles in the churches of the county, although the Perpendicular is the most common. Some of the churches bear traces of all the styles, but in many cases, they are not so composite. We can illustrate the composite character of the architecture by selecting Harrow Church for consideration. This church, dedicated to St Mary, is situated on the top of a hill, and is visible from a great part of the valley of the Thames, while from its tower may be commanded a wide prospect, including, it is said, portions of thirteen counties. From early Saxon times the manor of Harrow belonged to the archbishops of Canterbury, and this connection may account for some of the features of the church. A church, built here by Lanfranc, was consecrated by his successor, St Anselm, in 1094. As we have already seen, the tower and the west doorway belong to this period, while the north and south doorways have good thirteenth century mouldings. The columns and arches of the nave are Early English in style, and the chancel arch, lancet windows in the chancel, and the font are also of the same period. The Perpendicular work in this church may be seen in the roof of nave and transepts, the aisle windows, and the clerestory. Here then we have a Middlesex church combining work of at least three periods, together with modern additions.

One of the best examples of an Early English church is the small church at Littleton. This style is also to

Harrow Church, from the Cricket Ground

be found in the churches at East Bedfont, Cowley, Perivale, Stanwell, and Hayes. Of the Decorated style the nave of Northolt is a good example, and other work of this period may be seen at Harefield, Enfield, Harlington, and Ickenham. As we have already remarked, Perpendicular work is very abundant throughout the county. The towers of the churches at Chiswick, West

Littleton Church

Drayton, Edgware, Pinner are all favourable examples of the Perpendicular style, and much good work of the same period is found at South Mimms, Ruislip, Hendon, Stanwell, Tottenham, and Uxbridge. The Middlesex

Harmondsworth Church

church towers, especially those at the west end, are the special features of the fifteenth century, and they were no doubt so generally built as it was then the desire of the folk of almost every parish to possess a ring of bells.

They are mostly of stone or rubble, or of flints with stone facings, and they have usually a projecting turret at one of the angles, which rises a few feet higher than the rest of the battlements. Heston is one of the best examples of this construction. In some cases, as at Harlington, Harmondsworth, and West Drayton, the turrets are capped by cupolas, but these are mostly of later date. When this desire for bell-towers became general, the smaller and poorer parishes turned to timber as a substitute. At Greenford and Perivale are two wooden towers built up as additions at the west end from the ground. Other wooden towers are at Cowley, Kingsbury, and Northolt, where a short wooden spire surmounts the tower.

The almost total absence of screens in the Middlesex churches is very noticeable. When we remember that the county, though destitute of stone, had formerly an abundance of timber, we are somewhat surprised; but the reason seems to be that the smallness of the early buildings led to such a rebuilding of the churches that almost the whole of the old fittings disappeared. No rood-screen is now to be found in the county, although remains are to be seen at Cowley and Hayes; while Ruislip has some survival of the rood-loft, and South Mimms has a most beautiful parclose.

Middlesex has a poor display of fonts, though they are of varied periods and of considerable interest in a few cases. There are good Norman fonts at Hendon, Hayes, and Harlington, but the most noteworthy is the Perpendicular font at West Drayton.

The timber work in the roofs of Middlesex churches suffers in comparison with the elaborate work in the East Anglian churches. In some of the churches, such as East Bedfont, Cowley, and Perivale, the naves retain the tie-beam and king-post roofs dating from the fifteenth century, while a good deal of the old cradle-roof, with bosses at the intersection of the timbers, is remaining at Hayes.

Of early monuments in the Middlesex churches there are very few, and this is probably accounted for by the preference of the old county families to be buried in the larger churches of London. There are, however, a good many brasses, although a large number have disappeared during recent restorations. The best is that of Lady Tiptoft at Enfield, but those at Harrow and Hillingdon are noteworthy. At Harrow, the earliest is that of Sir Edward Flambard, who died in the reign of Edward III, but that which usually attracts most attention commemorates John Lyon, the founder of Harrow School, who died in 1592. The later monuments in the Middlesex churches chiefly date from the early part of the seventeenth century. They are usually kneeling figures of painted alabaster, and Nicholas Stone, a notable sculptor, has left fine examples of this work in the churches of Cranford, Enfield, Stanmore, and elsewhere.

The old glass in the windows has been nearly all destroyed, but some interesting fragments are to be found at Greenford and South Mimms. Some of the churches had wall-paintings, but these have all disappeared except at East Bedfont, Ruislip, and Hayes.

The Tiptoft Brass
(*At Enfield Church*)

A great change has come over the churches of Middlesex. Fifty or sixty years ago their condition was such that a writer spoke of their "disgraceful state of neglect and dilapidation." Now they are in good order and well-equipped for public worship, while the churchyards have been described as "models of loving care and neatness." Here we may note that Hayes and Ruislip are the only two churches having noteworthy lych-gates at the entrance to their churchyards.

Before the Reformation, the religious houses in Middlesex, outside the modern county of London, were not numerous. The monastery of Syon, founded in 1414 by Henry V, was the most remarkable convent of that period in England. It was dedicated to St Bridget, and the present Syon House at Isleworth stands on its site. Berrymead Priory at Acton is said to have been built on the site of a convent. Friern Barnet, as its name implies, belonged to a religious order, the Knights of St John of Jerusalem, but the remains of the Friary have long since disappeared. Near Edgware Church, which was probably part of a monastery of St John of Jerusalem, used to stand a house of refreshment for monks journeying between St Albans and London. At Harmondsworth there was a Benedictine monastery of the Holy Trinity, and at Bentley Priory, Stanmore, a house of the Austin Canons. Before the Order of the Knights of St John of Jerusalem was suppressed, Cardinal Wolsey obtained a lease of their property from the last prior at Hampton. A Priory of Brethren of the Holy Trinity stood at Hounslow, but nothing of the original building is in existence.

16. Architecture (*b*) Domestic—Hampton Court Palace.

Unlike the neighbouring county of Essex with its fine Norman keeps at Colchester and Hedingham and ruins of castles at a dozen other places, Middlesex has nothing to show in the way of military architecture. This is probably owing to the fact that William the Conqueror considered the Tower of London, which was then in Middlesex, to be sufficient, and he and his successors did not give any licences to crenellate in the metropolitan county. Middlesex, too, is without moated houses, though moats are to be seen at Ruislip, Northolt, and a few other places; and it has no famous seats such as Audley End in Essex, Hatfield House in Hertfordshire, or Knole House in Kent. The great nobles, who had fine town houses in London, preferred to live outside the metropolitan county, and so we shall find that the domestic architecture of Middlesex is somewhat restricted in its range. Hampton Court Palace and the Jacobean mansion at Swakeleys are the best examples in the county of the earlier period. The modern houses, in many cases with magnificent interiors, are devoid of external architectural distinction. We shall glance at some of them in the next chapter, and later, note will be made of the literary associations connected with Pope's Villa at Twickenham and Horace Walpole's Strawberry Hill.

In devoting this chapter to Hampton Court Palace, it may be pointed out that in the Tudor period, to which this palace belongs, the houses of the nobles were built less like fortresses, and more as comfortable homes for the owner, his family, and his servants. An authority on architecture has remarked that "under the Tudors the Gothic style went out in a blaze of glory," and that "Wolsey was the last professor of Gothic." Even in Wolsey's work we see the beginning of a new influence, and Italian details frequently remind us that the Renascence architecture was developing. The general plan of the great Tudor architects was to build a house round a quadrangle having the hall in the middle, and a wing on either side. The characteristics of the building were the quaint gables, mullioned windows, and large chimneys. The hall was often supported by massive beams of oak; the lofty rooms had high wainscoting, ornamental plaster ceilings, and beautiful chimney-pieces. The noble oak staircase with its carved balusters and heavy hand-rail would also be a conspicuous feature in such a building.

We are now prepared to consider Hampton Court, the palace of Wolsey and afterwards of Henry VIII, then of all our sovereigns in succession to George II, and now, by royal favour, a palace free to the enjoyment of all. Hampton was fixed on by Wolsey as the site for the palace because it was considered one of the healthiest and pleasantest spots in the south of England, and at a convenient distance from the metropolis. When he was at the height of his power he obtained a lease of the manor and manor-house from the prior of St John, and in 1516 commenced the erection of a magnificent mansion,

Hampton Court Palace

which he furnished in a style of corresponding splendour. A contemporary writer says of this palace: "One has to traverse eight rooms before you reach the cardinal's audience chamber, and they are hung with tapestry which is changed once a week." Before the palace was completed, however, in 1526, Wolsey found it advisable to present it to his royal master. Henry VIII completed it

Hampton Court Palace: Base Court

according to the design of Wolsey's architect, and soon afterwards the king enclosed a large tract of land in the neighbourhood, which thus became a Royal Chase. We have already hinted at the subsequent history of the palace which played an important part in the history of England, and so we may now proceed to a brief study of the structure.

Wolsey's palace consisted of five great courts, surrounded by public and private rooms of great dignity and splendour. There are now but three of these courts, for Sir Christopher Wren demolished two of Wolsey's courts and remodelled a third. The first quadrangle at the western entrance alone remains as originally erected by Wolsey, and is entered by a fine gatehouse. The chimney-shafts throughout the buildings are worth careful attention, and the tall gatehouse with its handsome oriel is very striking. The gateway leads to the second quadrangle called the Clock Court from a curious astronomical clock in the highest storey of the tower. On the north side of this court is the Great Hall, built by Henry VIII, on the site of Wolsey's Hall. It is of noble proportions, and entering it from beneath the Minstrels' Gallery, the effect is very striking. Wide Tudor windows are on both sides of the hall; on the walls beneath hang tapestries; and above all there is the grand open hammer-beam roof, which was restored in 1820. The windows are filled with heraldic blazonings, and the tapestries represent the principal events in the life of Abraham. Beyond the Hall is the Presence Chamber, hung with tapestries and a series of seven cartoons in monochrome. The chapel is small and has a good groined roof.

The State Apartments are entered under the colonnade at the south-east corner of the Clock Court. The many rooms vary greatly in size, but generally they are characteristic of the palatial architecture of the time. The carvings were the work of Grinling Gibbons, and most of the rooms contain furniture and upholstery of

The Great Hall, Hampton Court Palace

the time of William III, Anne, or George I. The Queen's Gallery is ornamented with Gobelin tapestry, and the Long Gallery has the Cartoons of Raphael. The King's Staircase, by which the State Apartments are reached, is one of the best examples in England of the "grand staircase" of the Louis XIV period. The

The Canal, Hampton Court

mural decorations and the paintings by Verrio are some of the characteristics of this staircase.

The chief attraction of the State Apartments is the collection of pictures, about 1000 in number, contained in them. Many are of great value and are portraits of notabilities by such painters as Holbein, Kneller, Lely, and other contemporary artists.

The charming gardens owe their general form to

Charles II, though they were extended and remodelled by William III and Mary. The grounds have been altered, but much of their formal trimness has been retained. The canal, with its avenue of lime-trees, three-quarters of a mile long, is one of William's devices, and the oval basin with its fountain and gold fish are of the same period. The noted vine dates from 1769; and the Maze, alluded to elsewhere, is the most popular spot in the grounds with holiday visitors. North of the palace is the Wilderness, always interesting, but seen at its best in spring; while beyond the lawns and gardens on the east stretches the Home Park, which is eclipsed by the larger Bushey Park to the north. It will thus be seen that Wolsey's palace is placed amid the most beautiful surroundings, and is one of the most popular holiday resorts within easy reach of London.

17. Architecture (*c*) Domestic—Famous Seats, Cottages.

Swakeley House or Swakeleys is the best example of a Jacobean mansion in Middlesex. The house is a spacious, picturesque, red-brick mansion dating from 1638. Although it has been little altered externally, it is in good preservation. It has been remarked that "the style is wanting in the playful exuberance of the true Elizabethan or Jacobean, but it is stately and effective; and among the majestic elms which surround it, and with the old church by its side, Swakeley forms an excellent representative old English manor-house." An enriched doorway is in the principal front which is carried up four storeys high; and the projecting wings terminate in large bay windows. The gables are numerous and irregular, and above are stacks of ornamented chimney shafts. Pepys in his *Diary* gives us a picture of the house when it belonged to Sir Robert Vyner, who lived in this house in 1665. He writes: "And so we together merrily to *Swakeley*, to Sir R. Vyner's; a very pleasant place....He took us up and down with great respect, and showed us all his house and grounds; and it is a place not very modern in the garden nor house, but the most uniform in all that ever I saw....The window-cases, door-cases, and chimneys of all the house are marble....After dinner, Sir Robert led us up to the long gallery, very fine, above stairs, and better, or such, furniture I never did see."

Syon, or Sion House, is the seat of the Duke of Northumberland, and stands in a small but pretty park which stretches from Brentford to Isleworth along the left bank of the Thames and opposite Kew Gardens. The present house, occupying the site of Syon Monastery, is a large quadrangular building, with a square tower at each corner. It is faced with Bath stone, and is crowned with an embattled parapet. In the middle of the west front there is an embattled portico which serves as the grand entrance. The east or river front has a projecting central bay carried the whole height of the building, and crowned with the well-known lion which formerly stood on the Strand front of Northumberland House in London. The grounds of Syon House, though level, are charming, and have magnificent trees. The gardens are of great extent and beauty, and here was formed by the Protector Somerset one of the first botanic gardens in England. The Great Conservatory, designed by Fowler, is said to contain the finest private collection of tropical plants in our country.

The present Chiswick House was built in the early

Syon House

part of the eighteenth century near the site of the former Jacobean mansion of the Earl of Somerset. Horace Walpole thus writes of this house:—"His lordship's house at Chiswick, the idea of which is borrowed from a well-known villa of Palladio...is a model of taste, though not without faults, some of which are occasioned by too strict adherence to rules and symmetry." Later in the eighteenth century, two wings designed by Wyatt for the Duke of Devonshire were added, and since then no material additions have been made to the house. Chiswick House was designed as a summer villa, and the garden and grounds were treated as part of the design, being lavishly decorated with urns, obelisks, and sculpture. Walpole wrote: "The garden is in the Italian taste,

Chiswick House

but divested of conceits"; and in these grounds open-air entertainments were given to monarchs and great people down to the last century. King Edward VII, when Prince of Wales, occupied Chiswick House for some time, but of late it has been in private hands.

Osterley House, which stands about two miles from Brentford, stands on the site of an old mansion built by

Gresham in the sixteenth century. The present mansion was built from designs by Robert Adam in the latter part of the eighteenth century, and is nearly square in plan with turrets at the angles. It is of red brick, and the Ionic portico in the centre of the principal front is approached by a flight of steps. The interior is splendid, and contains some antique statuary and interesting pictures. The great hall with the fine staircase and ceiling was painted by Rubens. The house stands in a park of 350 acres which has two lakes and much good timber.

Wrotham Hall, in South Mimms, was built for Admiral Byng in 1754, only a few years before his execution. The house is a spacious, stately, semi-classic structure of the Georgian period, and consists of a centre and wings with a recessed portico.

Hillingdon House is a large plain mansion of two storeys erected by the Duke of Schomberg in 1717. It stands on a slight elevation, in an undulating and richly wooded park. Charles Greville, the diarist, wrote of it:—"I never go to that place without looking with envy and admiration at the scene of so much happiness." Gunnersbury is one of the finest mansions in the neighbourhood of London but is much later, the present house having been built in 1801.

At Harmondsworth there is a very remarkable old barn, probably of monastic origin. It is of great size, being 191 feet long and 38 feet wide, and is divided into three floors. The walls are of pudding-stone, and the open roof, of massive oak, is an excellent example of old timber-work. The body of the barn is divided into a

Ruislip

nave and aisle, by two rows of oak pillars of immense thickness, which rest on square blocks of sandstone. It is still filled with wheat and other corn grown on the farm, and is thus a notable and rare survival in Middlesex.

In closing this chapter there is little to say about the distinctive character of the cottage architecture of the county. In no way may it be compared with that of Surrey, Kent, or Essex, where notwithstanding the great changes of recent years, there are yet many good examples of cottages dating from the seventeenth and eighteenth centuries. Middlesex has lost all its typical rusticity, and perhaps Ruislip is the only village which has kept its pretty old village street with its irregular gabled houses. It is now proposed to convert Ruislip into a garden city, so that even this old-world village will soon be nothing more than an annexe of London. The suburban houses which have sprung up in recent years need no description. They are generally mean and monotonous and have been built with only one idea—to place as many houses as possible on a certain site at the least possible expense. We almost wish for the revival of a certain statute of Elizabeth's reign which required that any cottage or building constructed for habitation should have four acres of land attached to it. Perhaps then we could realise what Morris pictures in *The Earthly Paradise*,

> "And dream of London, small, and white, and clean,
> The clear Thames bordered by its gardens green."

18. Communications, Past and Present —Roads, Railways, Canals.

There is no doubt that the Trinobantes, who lived in Middlesex when Caesar invaded our land, made paths or trackways through the primeval forest which then covered their territory. These British trackways are in many instances the origin of parish roads, for they led from one settlement to another. In some localities, where high roads have been constructed, portions of these trackways remain; and when they have not been metalled by the local authorities they are known as grass roads. Although British trackways have been traced in various parts of Middlesex, they have long since ceased to be of importance as main roads, and we only refer to them as they were prior to the fine roads made by the Romans.

The roads constructed during the Roman occupation were part of the network of roads that covered the Roman world, and for many centuries they continued to be the chief means of communication within our island. Some of them are still in almost perfect condition, while others form part of the foundation of roads now in use. The course of these roads was planned with skill, and with a knowledge of the general features of the district; but we need not suppose that they were constructed by the same master nor at the same time. When we consider the extent and permanent character of them, we may say that the Roman roads claim a foremost place among the remains of the Romans in our country.

In Middlesex the Roman roads are of great importance and give access to all parts of the county. The main object of the Romans in making these roads was undoubtedly military, and the line of communications between their various forts and stations was carefully guarded. These roads were not only used as great

Highgate Archway

channels of communication, but they often served as the limits of the various divisions of their conquests, while the boundaries were marked by mounds, stones, or trees. It is well known that the ridges of Roman roads were often made the boundary between parishes and townships; and for many miles boundaries sometimes follow roads which are certainly Roman. Thus, in

Middlesex, parish boundaries follow Watling Street along the Edgware Road continuously for five miles from Oxford Street to the river Brent, and again after an interval for a further distance of two miles.

The chief Roman highways are now known to us for the most part by the names given to them by our Anglo-Saxon forefathers. Most of the great Roman roads converge on London, and coincide in a remarkable manner with our modern railroad communication.

The best known and most important of the Roman roads through Middlesex was Watling Street. Beginning at Dover, it passed through the north of Kent and entered London. From London it went in a north-westerly direction to Chester and through Lancashire. Even to-day, a portion of this road is called Watling Street in the City of London, and its course through London is probably by way of Holborn and Oxford Street. The first certain trace of it is at Tyburn, and from that point, Edgware Road and its continuation occupy the line of Watling Street to Brockley Hill, rather more than 10 miles distant. From Brockley Hill, which was near the Roman *Sulloniacae*, it proceeds through St Albans, and a branch is supposed to go to Watford and Tring. A short time ago, the Roman paving was cut through in the Edgware Road, and was found to consist of large black nodular flints, weighing from four to seven pounds each, on a bed of rammed reddish-brown gravel of varying thickness. The gravel was supported by dwarf walls of gravel concrete, and on the levelled surface of the gravel, lime grouting appears to have been laid, in which the flints were firmly set. The workmen found that it gave them much more trouble to break up than the modern concrete road above.

The second important Roman road through Middlesex is that from London to Staines through East Bedfont. This road made a junction with Watling Street at the south end of Edgware Road and then continued along the line of the present Bayswater Road, almost in a direct course, passing through Brentford and Hounslow to Staines—the Roman *Pontes*—on the Thames.

The third Roman road in Middlesex is that from London known as Ermine Street. It follows roughly the line of the present Kingsland Road and its continuation through Stoke Newington and Tottenham to Edmonton, and thence goes straight on to Forty Hill and Maiden Bridge. Near where Shoreditch church stands, Ermine Street was crossed by the Roman road passing to the north of London in the line of Old Street, from Tyburn to Old Ford, and was then continued over the Lea into Essex.

The last Roman road in Middlesex we will mention is that from Cripplegate in London. It passed thence to Stoke Newington, Hornsey, and on to Enfield Chase, along the modern road to Southgate and Potter's Bar.

We must now leave the times of the Romans and come down to a more recent period. At the present time, the communications of Middlesex are very good, owing to the radiation of great roads and railways from London to all parts of the east, west, and north of our country. The great Middlesex highways are kept in excellent condition, and form a striking contrast to their almost impassable and insecure state in the eighteenth century. Then the high roads were often mere tracks, and were beset by highwaymen, so that travellers often passed along bridle ways through fields, where gibbets warned them of the perils of the district.

The great roads through Middlesex radiating from London are six in number. Watling Street or Edgware Road follows the Roman road already described. The Exeter Road also follows the Roman road through Hounslow to Staines, and the Ware Road is practically Ermine Street through Edmonton and Enfield to the county boundary. The Great North Road dates from 1386, and runs over Highgate Hill, across Finchley to Whetstone, and thence to Barnet. The Oxford Road runs through Ealing and Uxbridge, and was of much importance in the coaching days, when it was considered a great feat to travel from London to Oxford in 3 hours 40 minutes. The Bath Road goes from London through Hammersmith, Brentford, and Hounslow to the west. This was a most important thoroughfare in the eighteenth century, when Bath was the resort of London society and fashion. The coaches ran from Piccadilly and did the journey in about 16 hours in 1784. This was considered very rapid, but in 1837 the Bristol Mail covered the distance in less than 12 hours with 14 changes of horses.

When the coaches on the main roads had reached their highest efficiency, railways were gradually supplanting them; and in 1838 the portion of the Great Western

Railway from London to Slough was opened. The county is now everywhere crossed by railways, which have their termini in London. Besides the great trunk railways, Middlesex is also served by various underground and tube lines to Finsbury Park, Highgate, Harrow, and other places that were considered rural a few years ago.

The G.W.R. Viaduct at Hanwell

Within the present century the means of communication have been still further increased by the construction of electric tramways from the county boundary to the outlying districts such as Hounslow, Uxbridge, Sudbury, and Edgware. The result is that, owing to the cheap and varied means of transit, large districts which not long ago were well-timbered parks and open spaces, are now covered with small villas and cottages; and even Ruislip, which was considered outside the range of speculative builders, is likely to become the largest garden city in England.

We may close this chapter on communications by glancing at the canals of the county. They were constructed just before the advent of railways, and for some

Harrow, from the Railway

years were very prosperous. Now many of them are neglected or derelict, owing to the quicker means of communication and transit by which they have been superseded. The Grand Junction Canal, opened in 1805, gives communication between the manufacturing centres of the midlands and the metropolis, and follows the Colne valley from Watford to West Drayton, thence turns east to Hanwell on the Brent, and afterwards south to Brentford. The Paddington Canal branches from the Grand Junction Canal at Bull's Bridge on the Crane, and passes through the central part of the county to Paddington, whence it is continued as the Regent's canal round the north of London to the Thames at Limehouse. In looking back over the early history of the formation of canals, we find that the country was in a perfect ferment about canal shares in the last decade of the eighteenth century, and all kinds of worthless and speculative schemes were set on foot, which brought ruin to many investors. By the end of the eighteenth century more than 2000 miles of canals were opened for traffic in England, and most of the large towns were accommodated with the means of easy transport for their goods to the principal markets. At one time, in 1806, during a period of great excitement, it was contemplated to despatch troops from London for Ireland via the Paddington Canal. The following curious paragraph is taken from the *Times* of December 19, 1806. "The first division of the troops that are to proceed by Paddington Canal for Liverpool, and thence by transports for Dublin, will leave Paddington to-day, and will be followed by others to-morrow, and on Sunday. By this mode of conveyance the men will be *only seven days* in reaching Liverpool, and with comparatively little fatigue, and it would take them above fourteen days to march that distance. Relays of fresh horses for the canal-boats have been ordered to be in readiness at all the stations."

19. Administration and Divisions—Ancient and Modern.

In order to get a just idea of the present administration of the county of Middlesex we must remember that many of our existing institutions can be traced back for a thousand years or more; and, although times have changed since our Saxon forefathers divided the land into self-governed divisions, we may look back with much satisfaction on the gradual steps by which our present mode of local government has been evolved from the methods of earlier times.

The government of each county in Saxon times was partly central and partly local. The central administration was from the county town, and the local administration was carried on in the hundreds, and parishes, and manors. The chief court of Middlesex in the earliest times was the shire-moot. It met twice a year, and its two chief officers were the Ealdorman and the Sheriff (shire-reeve), the last of whom was appointed by the King. The shire-moot was the survival of an earlier gathering, the folk-moot, which was held often in the open air on a haunted mound or round some aged oak of sacred memories. To the shire-moot were sent representatives of each rural township and of each hundred. The sheriff published the royal writs, assessed the taxation of each district, and listened to appeals for justice.

Middlesex is still divided into six hundreds, even as it probably was in the days of the Saxons, and later still in the Domesday Book. These hundreds are Ossulston (Oswald's town) which includes London ; Isleworth and Spelthorne along the Thames; Gore and Elthorne on the west, and Enfield on the east. Each hundred probably consisted at first of one hundred free families, and had its own court, the hundred-court, which met every month for business. Each hundred was sub-divided into townships, or parishes as we now call them; and every township had its own *gemot*, or assembly, where every freeman could appear and help to make laws for the township and appoint officers to see that they were enforced. The town-moot was held whenever necessary; its chief officers were the reeve, who acted as president, and the tithing-man, who was a constable, something like our policeman of to-day.

There was thus a kind of triple arrangement for the government of each county; but besides these courts of the shire, the hundred, and the township, there were also courts of the manor. The manors were holdings of land which varied in extent, some being as large as the township, while in many cases they were much smaller. The manors were under the lords of the manor, and were held on various conditions, such as rendering service, or homage to the King. The manors had their own courts, such as court-leet, court-baron, and customary-court. In these courts the lord and his tenants met and settled the affairs of the manor, such as those relating to the common fields, the right of enclosure, and the holding of fairs and markets. These manor-courts are still held in many parts of Middlesex, and although they have lost nearly all their former importance, it is of interest to remember that in them we have survivals of the work of our forefathers more than one thousand years ago.

We may now consider the present method of county government in Middlesex. The Lord-Lieutenant and the High Sheriff are the chief county officers. The former is generally a noblemen or a large landowner, and is appointed by the Crown; the Duke of Bedford is the present Lord-Lieutenant of Middlesex. The Sheriff is chosen every year on "the morrow of St Martin's Day," November 12. The County Council is now the central authority and conducts the main business of the county. It was constituted in 1889 and holds its meetings at the Middlesex Guildhall, Westminster. The Middlesex County Council consists of 19 Aldermen and 59 Councillors, the latter being elected by the ratepayers, while the former are co-opted by the councillors. The keeping of roads and bridges in good repair, the regulations with regard to lunatic asylums, and the carrying out and enforcing of laws passed by Parliament are some of the important duties of the County Council.

For local government in towns and parishes, an Act was passed in 1894, when new names were given to the various administrative bodies which had been known as vestries, local boards, highway boards, etc. In the towns and larger parishes the chief governing authorities are now known as urban district councils, of which there are 33 in Middlesex. There are also four rural districts, and the other 21 smaller parishes have parish councils, while one has a parish meeting.

Middlesex has also four Poor Law Unions wholly, and parts of three others, each of which has a Board of Guardians, whose duty it is to manage the workhouses, and appoint various officers to carry on the work of relieving the poor and aged. Willesden is a separate Poor Law parish. Two other areas have yet to be men-

Harrow School: The Terrace

tioned, the municipal boroughs of Ealing and Hornsey. They were formerly urban districts, but received their charters from the Crown in 1901 and 1903.

Besides the Court of Quarter Sessions at Westminster, Middlesex has also one at Brentford, and eight Petty Sessional Divisions, each having magistrates or justices of the peace, whose duty it is to try minor cases and award punishments. The county lies within the jurisdiction of the Central Criminal Court and the Metropolitan Police.

We now pass to ecclesiastical administration, which has altered very little during the last thousand years; indeed the Church existed before the State and had its dioceses or divisions and its own courts. The northern

Mill Hill School

dioceses are under the care of the Archbishop of York, while the southern dioceses are under the Archbishop of Canterbury, the primate of all England. Middlesex is ecclesiastically in the province of Canterbury and the diocese of London, and forms an archdeaconry, which is subdivided into 14 rural deaneries and 153 ecclesiastical parishes or districts. These ecclesiastical districts are in

60 civil parishes within the borders of the county. At one time, the civil and ecclesiastical parish were one and the same, but as the population increased it became necessary to subdivide many of the civil parishes for church purposes.

The educational affairs of the county are administered by the Education Committee of the County Council, which has control of secondary and elementary education in the greater part of Middlesex. There are, however, separate elementary education committees for 13 of the largest towns and parishes.

Of the public schools in the county Harrow is of course the most famous (see pp. 151, 152); Highgate School, which dates from the sixteenth century, is referred to elsewhere. Mill Hill School was founded in 1807 for the education of Nonconformists and re-organised on a broader basis in 1869.

Middlesex prior to 1885 returned two members to Parliament, but since that date it has had seven members, representing the divisions of Enfield, Tottenham, Hornsey, Harrow, Ealing, Brentford, and Uxbridge.

20. Roll of Honour.

The proximity of Middlesex to London has done much to rob the county of its individuality, and we do not hear Middlesex people speak with pride of their native county as those born in Yorkshire or Devonshire. The oldest Middlesex families lived within the borders of the present Metropolis and were of exalted rank. Thus the Russells since the sixteenth, and the Cecils and Howards since the early seventeenth century had their town houses in London, but their country seats were in Bedfordshire, Hertfordshire, or Sussex. Although, however, there were no great Middlesex families in the same sense as in the neighbouring counties, there is scarcely a village in the county without its memories of some one who made himself famous in the neighbouring London.

From the time of Henry VIII to the eighteenth century Hampton Court was one of the royal residences. As we have already seen, Henry VIII added considerably to Wolsey's buildings, and in the latter part of his reign it became one of his principal residences. Edward VI was born at Hampton Court in 1537, and his mother, Jane Seymour, died there soon afterwards. Philip and Mary kept their Christmas at Hampton Court with great solemnity in 1558; and Queen Elizabeth after she came to the throne frequently resided at Hampton Court. Here it was that James I met the conference of divines in 1604, and Charles I was kept a prisoner till he made his escape on November 11, 1647. Charles II and James II lived occasionally at this palace, and William III spent a good deal of time here, engaging Sir Christopher Wren to improve and enlarge it. Queen Anne and the first two Georges lived here occasionally, but subsequently it ceased to be a royal residence. York House, Twickenham, was another royal residence, and the Princesses Mary and Anne were born there. Their father, James II, had some notoriety in Middlesex, for he gathered his forces on Hounslow Heath in 1686 to meet his enemy. William IV, when Duke of Clarence, lived at Bushey House for 36 years, and Orleans House, Twickenham, was for a time the centre of the French Royalists. King Edward VII, when Prince of Wales,

Cardinal Wolsey

occupied Chiswick House, and its beautiful grounds were much enjoyed by his young family.

Among the most noted divines connected with the county, we find that Grocyn, one of our earliest Greek scholars, was rector of Shepperton from 1504–1513.

The greatest churchman of Middlesex was undoubtedly Wolsey, who in 1515 bought the manor of Hampton. As Archbishop of York he was then in the plentitude of his power, and without delay and at a vast cost he proceeded to build so large and stately a palace that, according to Stow, "it excited much envy." The King himself asked Wolsey why he had built so costly a house, and received the ready and adroit reply, "To show how noble a palace a subject may offer to his sovereign." Thomas Fuller, author of the *Worthies of England*, was rector of Cranford for several years and was buried there in 1661. Richard Baxter, who wrote *The Saint's Everlasting Rest*, was living at Acton after the Restoration, and was committed to prison for holding services in his own house. Cardinal Newman was educated at Dr Nicholas's school at Ealing, and Cardinal Vaughan founded St Joseph's College, Mill Hill. Archbishop Temple began his long and useful career as first principal of Kneller Hall, which was founded as a training college for schoolmasters.

Among our statesmen, Middlesex has memories of Thomas Cromwell, Earl of Essex, who was the prime mover in the spoliation of the monasteries, and who lived at the manor-house, Brondesbury, for some time. William Lenthall, Speaker of the House of Commons during the Long Parliament, lived at Whitton, and Sir Matthew Hale, Lord Chief Justice in the stormy Stuart times, was a resident at Acton. Fox, the parliamentary opponent of Pitt, and George Canning, the prime minister of George IV, both died at Chiswick House, the former in 1806, and the latter in 1827. Lord Westbury, a

Victorian Lord Chancellor, lived at Highgate, and W. E. Gladstone was a frequent visitor at Dollis Hill. There is a memorial of him in the church at Willesden, and the park in that town is named Gladstone Park.

Middlesex can boast of no great man of action of the first rank. Sir Henry Lawrence, the Indian soldier, and Lord Lawrence both went to school at Ealing. It is really remarkable that Ealing educated so many great men, and we may here refer to its excellent school, which long enjoyed the reputation of being one of the first private boarding-schools in the country. In its best days it had about 300 scholars, among whom many have attained distinction in various walks of life. General Eliott, afterwards Lord Heathfield, and the defender of Gibraltar, lived at the Grange, Ealing.

When we come to the men connected with literature who have made their home in Middlesex, we have quite a wealth of names. Among historians, Richard Gough the antiquary, and editor of Camden's *Britannia*, and Bishop Thirlwall, historian of Greece, both lived in Enfield. Isaac Disraeli, author of the *Curiosities of Literature* and various historical works, and father of Benjamin Disraeli, the future prime minister, was born at Enfield in 1766, and resided here till 1804.

Among the poets associated with Middlesex we find a goodly number, ranging from Suckling to Tennyson. Sir John Suckling, one of the Stuart poets, was born at Whitton in 1609. Andrew Marvell, the poet-panegyrist of Cromwell, lived in a cottage on Highgate Hill. A long, low, modest, wood and plaster building, with a central bay window and porch, it stood till 1869 when it was removed by Sir Sydney Waterlow. Of all the poets connected with this county, Pope undoubtedly holds the most important place. His villa has made Twickenham famous wherever English literature has reached. He took

Pope's Villa, Twickenham

a lease of this house with about five acres of ground in 1719, and lived here till his death in 1744. Pope was fond of his garden and proud of it, for, though of small size, he contrived with the aid of professional gardeners, to make it one of the prettiest gardens in England.

He was the first to break through the Dutch formality of the gardens at Hampton Court, and to revert to a more natural style. Pope's success in landscape-gardening was not due to a happy chance, but was the result of careful and serious consideration, as is evident from many of his letters. The favourite amusement of his declining years was the making of the grotto, which was formed by lining the tunnel under the Teddington road with shells, spars, and minerals, liberally given by his friends. Coleridge, the poet, went in 1816 to the house of the Gillmans at Highgate, where he spent the last years of his life. Mr Gillman was a surgeon, and Coleridge went to reside with him to be under his surveillance in order to break himself of the habit of opium eating. Here it was that Carlyle visited him, and wrote that "Coleridge sat on the brow of Highgate, in those years, looking down on London and its smoke tumult, like a sage escaped from the uncertainty of life's battle, attracting towards him the thoughts of innumerable brave souls still engaged there." Coleridge was buried in a vault in the graveyard of the old church at Highgate. On his monument is the inscription : *Poet. Philosopher. Theologian. He quitted the body of this death July* 25, 1834. Shelley went to school at Sion House, Brentford, and Samuel Rogers, the wealthy banker and poet-satirist, lived at Hornsey. Keats was educated at a school kept by John Clarke at Enfield, and lived with his brother at Well Walk, Hampstead. When not at school, the boy lived from his tenth to his fifteenth year in his grandmother's house, and was then apprenticed to an Edmonton surgeon named Hammond. Tennyson lived in Montpelier Row, Twickenham, and his son Hallam was baptized

The "Peachey Stone" and Byron's Elm, Harrow

at the church of that parish in 1852. In the roll of great men who were educated at Harrow, Byron is one of the most remarkable. He spent several years under

Dr Drury, and we have already referred to his favourite spot in Harrow churchyard. He wrote some verses "On a distant view of the village and school of Harrow on the Hill"—

> "Again I behold where for hours I have ponder'd,
> As reclining at eve on yon tombstone I lay;
> Or round the steep brow of the churchyard I wander'd,
> To catch the last gleam of the sun's setting ray."

We now pass from the poets to the men of letters who have shed lustre on the county. First we may place Francis Bacon, who in 1592, when M.P. for Middlesex, entertained Queen Elizabeth at his large mansion at Twickenham. After he had sold it, and not long before his death, he expressed a wish that it should be repurchased for a residence for deserving persons to study in, "since I experimentally found the situation of that place much convenient for the trial of my philosophical conclusions." The story of Bacon's last days at Highgate is well known. He was living at Arundel House, when, on a snowy day in the spring of 1626, he ventured out to try an experiment. He caught a severe chill and was carried home, where he died a few days afterwards. Horace Walpole, one of the best known men of letters of the eighteenth century, is inseparably connected with Twickenham. When a boy he spent a summer with his tutor at this place, and during a long life he retained his early liking for it. In 1747 he bought a house, the famous "Gothic Castle," which he afterwards named Strawberry Hill. Here he lived surrounded by his books and articles of *vertu* which he delighted to collect, and here he was visited by most of the great people of his day. He had his own private printing-press, and became an authority on pictures, furniture, ivories and mosaics—in short he was a man of taste. For fifty years Walpole

Horace Walpole

lived at Strawberry Hill, and was employed during most of that time in improving the house, adding to his collections, and enjoying his two passions, the lilacs and nightingales in his grounds. Leigh Hunt, essayist and poet, prided himself on being a native of Middlesex, for he was born at Southgate in 1784, "a prime specimen of Middlesex," and with the "sweet air of antiquity about it." Charles Lamb is associated in Middlesex with Enfield, where he lived "in an odd-looking gambogish coloured house," from which he moved in 1833 to Church Street, Edmonton, where he died in 1834. He was buried in the churchyard of that parish, and there the most interesting monument is that erected to him, with its long poetical inscription by Cary. John Walter, the founder of the *Times*, was buried at Teddington, and there, too, Blackmore, the author of *Lorna Doone*, lived for some years and died. Laleham, the quiet little riverside parish on the Thames, has intimate associations with the Arnolds. Thomas Arnold lived here as a private tutor for nine years before he became head-master of Rugby. His most distinguished son, Matthew Arnold, was born at Laleham in 1822, and lies buried in its churchyard. William Watson wrote a poem in Arnold's memory, which begins:

> "Lulled by the Thames he sleeps"

and ends:

> "And nigh to where his bones abide,
> The Thames with its unruffled tide
> Seems like his genius typified—
> Its strength, its grace,
> Its lucid gleam, its sober pride,
> Its tranquil pace."

Thackeray and Bulwer-Lytton both went to school at Ealing, and William Howitt and his wife, Mary Howitt, passed many happy years at Highgate. William

Howitt wrote a most interesting book on *The Northern Heights of London*, which gives the history and associations of Highgate and the neighbourhood.

Thomas Henry Huxley

Among men of science we have already referred to Bacon. In the eighteenth century, Peter Collinson, naturalist and antiquary, lived at Mill Hill, where his famous gardens were visited by Linnaeus. Huxley was born at Ealing and educated at Ealing school, where his father was a master, and Faraday died within the precincts of Hampton Court in 1867. Paxton, when a day workman in the Gardens at Chiswick, was discovered by the Duke of Devonshire and afterwards became famous as the designer of the Crystal Palace. At Mill Hill, Sir Stamford Raffles spent the last year or two of his life, while at Greenford Dr Parkin first made aniline dyes, which have played such an important part in our textile industries.

In the domain of art, Middlesex can boast of not a few great names. Sir Godfrey Kneller, the great portrait painter, lived in the Upper Mall, Chiswick, before removing to Whitton near Twickenham, where he died in 1723. He was buried in the church of the last place, where there is a monument with an inscription by Pope to his memory. It is said that ten reigning sovereigns sat to Kneller for their portraits, besides most of the persons of importance in his day. Hogarth, the painter, engraver, and pictorial satirist of the middle of the eighteenth century, lived and died in 1764 at Chiswick. Hogarth House, where he lived, still stands not far from the church, where there is an altar tomb to the painter with an inscription by Garrick. Turner, our greatest landscape painter, lived in the Upper Mall, Chiswick, from 1805–1814, and afterwards for some years at Twickenham. David Garrick, the actor, was a resident at Hampton from 1754 to his death in 1779, and Sir Christopher Wren had a house on Hampton

Court Green, where he passed the last five years of his life. Handel was organist at Whitchurch from 1718 to 1721, and on the organ he composed *Esther*, two *Te Deums*, and various anthems.

There are many other names of note for which we can only spare a little space. Sir Hugh Myddelton, the projector of the New River, lived at Enfield for some years, and Sir Rowland Hill, of Penny Post fame, was educated at Bruce Grove, Tottenham. Izaak Walton, who wrote *The Compleat Angler*, was familiar with the river Lea from its source downwards. John Wilkes in the eighteenth century, and Sir Francis Burdett in the nineteenth century, both made themselves notorious as M.P.'s for Middlesex. The former was re-elected three times before he sat as M.P. for the county in 1774. Three Middlesex worthies of a different type were John Lyon, the founder of Harrow School, William Penn, the quaker and founder of Pennsylvania, who lived at Brentford, and Wilberforce, the philanthropist, who resided at Highwood Hill for a period of five years. Perhaps it is fitting that this chapter should close with a brief reference to the first of the three. John Lyon was a yeoman, who many years before his death, conceived the project of establishing a Free School in his native village. Scarcely anything is known of his life, but he was buried in 1592 in the nave of Harrow church, and on his brass there is a lengthy legend ending: "Prayse be to the Authour of all goodnesse, who maketh us myndful to follow his good example."

21. THE CHIEF TOWNS AND VILLAGES OF MIDDLESEX.

(The figures in brackets after each name give the population in 1911; the references at the end of the sections are to pages of the text.)

Acton (57,497), a suburb of London on the Uxbridge road, has grown rapidly in recent years owing to improved communications with the metropolis. The manor has belonged to the see of London from earliest times. The parish church is modern; the tower of the old church was pulled down in 1877. During the Commonwealth period, Acton seems to have been quite a Puritan stronghold. Philip Nye, a notorious Puritan minister held the living; Francis Rous, Speaker of the Little Parliament, resided here; and Sir Matthew Hale, the celebrated Lord Chief Justice, and Richard Baxter, the eminent Puritan writer, also lived at Acton. About the middle of the eighteenth century Acton Wells were in great repute; and the Assembly Room was a place of fashionable resort. In later years, Fielding the novelist, Sir Bulwer Lytton, and Percy Lindley, the distinguished botanist, lived in the vicinity. (pp. 7, 61, 73, 82, 85, 102, 133.)

Ashford (6763), two miles east of Staines, is written Exeforde in Domesday, from the little river Exe. The present handsome church was built in 1858 on the site of the old Norman building. The Welsh Charity School, of the Society of Ancient Britons, was founded in 1814. It is a picturesque building of a modified Elizabethan character, with tall dormer gables, clock-tower, and high red-tiled roof. It was opened by the Prince Consort in 1857, and is intended for children of Welsh parents, born within 12 miles of the Royal Exchange. (p. 61.)

Brentford (16,571), the reputed capital of Middlesex, is divided by the river Brent into Old and New Brentford. It lies on the left bank of the Thames opposite Kew Gardens, and is a market-town with brewery, soap-works, etc. Brentford is an ancient town, and was the scene of a Danish defeat in 1016, and of a Royalist victory in 1642. As the polling place for Middlesex, Brentford was the scene of serious rioting during the Wilkes agitation. There are many references to the town in our literature, but they are generally connected with its dirt and squalor. We must, however, note that it is also well known for "the two Kings of Brentford" in *The Rehearsal* and *The Task*. Brentford has no buildings of interest and the churches are modern. The stone bridge over the Thames, which connects Brentford and Kew, was opened by King Edward in 1902. (pp. 5, 22, 31, 66, 67, 69, 70, 76, 81, 121, 128, 143.)

Chiswick (38,697), in a great loop of the Thames, is a suburb of London on the road to Brentford. It has breweries and coal-wharves, and launch and torpedo-boat works, as well as extensive market-gardens for the supply of the Metropolis. Chiswick Ait, or Eyot, is the first on the Thames above London. The river front known as the Mall extends from Chiswick to Hammersmith, and is a most interesting part of the neighbourhood. Chiswick House is a fine Palladian building belonging to the Duke of Devonshire. Here it was in 1806 that Fox died, and in 1827 that Canning passed away. The church stands quite close to the river and has a tower of flint and stone dating from 1435. It has many good monuments, and in the churchyard is the altar tomb of William Hogarth. The celebrated painter lived in an old-fashioned red-brick house which still stands not far from the church and is known as Hogarth House. (pp. 7, 13, 27, 69, 70, 96, 112, 132, 142.)

Cowley (1021), a pretty village between Uxbridge and West Drayton, stands on the Colne and Grand Junction Canal. Its church is a small and interesting pre-Norman building, and has Early English and Decorated work. (pp. 93, 96, 100.)

Cranford (615) stands on the Cran or Crane, three miles north-west of Hounslow. It is pleasantly situated on the Bath Road, and its church of Norman foundation has some interesting monuments. (p. 100.)

Ealing (61,222) was incorporated as a borough in 1901, and has developed owing to the railways and the electric trams. The churches are all modern, and the Town Hall was opened in 1888 by King Edward VII when Prince of Wales. The borough is well supplied with open spaces, such as Ealing Common, Walpole Park, and Lammas Park. Among the eminent men who have lived in Ealing may be named Bishop Beveridge, Dr John Owen, the Puritan Divine, and Huxley, who was born at Ealing in a house near St Mary's Church. This house was once a school of some repute, and many of its pupils rose to eminence. Cardinal Manning, Thackeray, Marryat, Sir Henry Lawrence, and Lord Lawrence are a few of the distinguished men educated at Ealing School. (pp. 7, 50, 63, 130, 134, 140, 142.)

East Bedfont (2426) is on the Staines road, three miles beyond Hounslow. The church is of great interest for it has a Norman south door, lancet windows in the chancel, and Perpendicular windows in the west and south sides of the nave. The churchyard has two yews that have been laboriously clipped and trained, so as to be famous examples of the art of the topiarist. In the neighbourhood, many Roman coins and relics have been found. (pp. 90, 91, 93, 96, 100.)

Edgware (1233), eight miles from Hyde Park Corner on the ancient Watling Street, was of some importance in the old coaching-days. The neighbouring palace of Canons was one of the great houses in the county; and the blacksmith's shop in which worked the musical William Powell stood on the west side of the main street. According to tradition, Handel once took shelter in this smithy, and Powell's performance on the anvil suggested to the great musician the well-known melody the "Harmonious Blacksmith." A short distance beyond Edgware is Brockley Hill, which is supposed to be near the site of the Roman station *Sulloniacae*. (pp. 96, 102.)

Edmonton (64,797) is a thickly populated, working-class district on the road to Ware. The parish church dates from the fourteenth century, and has traces of the original Norman building. Charles Lamb and his sister, Mary, are buried in the churchyard, and there are other literary associations of interest. The "Bell" at Edmonton has been rendered famous by Cowper's *John Gilpin*: and the town attained an earlier celebrity from being the locale of two plays of some note, *The Merry Devil of Edmonton*, and *The Witch of Edmonton*. (pp. 7, 32, 47, 69, 70, 73, 93, 121, 140.)

Enfield (56,338) is a large parish with a rapidly-growing artisan population. Enfield belonged to the Crown till the time of James I and has remains of an ancient Royal palace. Enfield Chase was a hunting-ground of nearly 4000 acres and was disafforested in 1777. Various parts of Enfield are known as Enfield Town, Enfield Highway, Enfield Lock, and Enfield Wash. Enfield Lock is the seat of the Government Small-Arms Factory, and gave its name to the once famous Enfield Rifle. The parish church, dating from the thirteenth century, is mainly a Perpendicular building of flint and stone. It has been much restored, but has some good monuments, especially a fine brass effigy of Lady Tiptoft, who died in 1446. Enfield is associated with the names of several eminent men. Isaac Disraeli was born here; Keats the poet was educated here; and Charles Lamb resided for many years at Chase Side. Writing of this house to a friend, Lamb says, "I am settled for life, I hope, at Enfield. I have taken the prettiest, compactest house I ever saw." He and his sister, however, soon removed to another house in Enfield, where they stayed till their removal to Edmonton. Lamb's favourite walks at Enfield were to the top of Forty Hill, and along the Green Lanes. (pp. 7, 49, 63, 70, 73, 91, 96, 100, 121, 130, 134, 140.)

Feltham (5135), a large straggling parish about three miles south-west of Hounslow, is somewhat uninteresting. Its church is modern, with some monuments from the older building. There are nursery and market-gardens in the parish, and the London County Council Industrial School is a great building for about 1000 boys.

Finchley (39,419) is one of the many Middlesex parishes that have lost their rural character during the last few years. Finchley Common formerly comprised more than 2000 acres, and as late as 1810 it was said to be in an uncultivated state. It shared with Hounslow Heath the distinction of being the favourite hunting-ground of highwaymen and was reputed unsafe for travellers even at the close of the eighteenth century. The Common was often used for military encampments, and played some part in our history, for here General Monk stopped on his famous march from the north in 1660, and a camp was formed here when the Pretender invaded England in 1745. The parish church, dedicated to St Mary, is a Perpendicular building, but so drastically restored that little of the original is left. (pp. 17, 42, 50, 61, 121.)

Friern Barnet (14,924) belonged to a religious order, the Knights of St John of Jerusalem, and the church has a Norman doorway, an Early English nave, and Decorated windows. The remains of the old Friary have disappeared, but the low red-brick almshouses, dating from 1612, have been restored and are picturesque buildings. (p. 102.)

Hampton (9220), on the left bank of the Thames, is 13 miles west of London. It is spelt Hamntone in Domesday, and the manor came in 1211 into the possession of the Knights of St John of Jerusalem, who in 1515 gave a lease of it to Cardinal Wolsey. Here he built the palace with such magnificence that he was obliged to present it to his master, Henry VIII. Henry and his successors down to George II lived much at Hampton Court, but in later times the private rooms have been the homes of royal pensioners. The State Apartments are good specimens of the palatial architecture of the time, and contain a collection of about 1000 pictures. The Gardens and Bushey Park are great attractions to Londoners, and the chestnut avenue in the Park is one of the finest in England. The parish church of Hampton was built in 1830, and took the place of an ancient structure. In this parish there are large waterworks under the control of the Metropolitan Water Board. Among the residences is Hampton House, which was the home of David Garrick. Here he was visited by Dr Johnson, who on leaving remarked, "Ah, David, it is the leaving of such places that makes a death-bed terrible." Sir Christopher Wren also lived in a house on Hampton Court Green, where he died in 1723. (pp. 25, 47, 80, 81, 103—110, 131, 142.)

Hampton Wick (2417) is opposite Kingston, with which it is connected by an important bridge.

Hanwell (19,129), an ancient parish on the Uxbridge Road, was given to Westminster Abbey by King Edgar in 959. Its population has recently increased owing to the coming of the electric trams. The church from designs by Gilbert Scott is modern, and the chancel has been decorated as a work of love by Mr Yeames, R.A. The Hanwell Lunatic Asylum accommodating 2500 patients is really in the parish of Norwood. The neighbourhood of Hanwell is green and pleasant, with the Brent winding through it. (p. 43.)

Harefield (2402), between Uxbridge and Rickmansworth, is in the most beautiful part of Middlesex. The river Colne and the Grand Junction Canal run along the west side of the parish; and the place has about it much quiet sylvan beauty. The village stretches for some miles along the road, and the church stands in

The Fourth Form Room, Harrow School

the grounds of Harefield Place. The church, of the Decorated period, is one of the most noteworthy in the county. Dating from about 1300 it has many interesting monuments; especially of the Newdegate family, the earliest brass being to Edetha Newdegate, 1444. There are some good houses and a few picturesque cottages in this parish, and the large Asbestos Works are on the canal by the lock. (pp. 10, 38, 40, 45, 46, 58, 69, 92, 196.)

Harrow-on-the-Hill (17,074) is one of the most picturesquely-situated places in Middlesex, and is famous for its church, its hill and the prospects from it, and above all for its school. Harrow Hill rises, abrupt and isolated, 400 feet high, and is a conspicuous and pleasing feature in the landscape for many miles on every side. The town occupies the crest and follows the slopes of the hill, while the school dominates it and colours it. In Domesday Book the name is written *Herges*; an

Harrow School: "Ducker"

early Latin form is *Herga super Montem*; and in 1398 it appears as *Harewe at Hill*. The manor belonged to the Archbishops of Canterbury long before the Conquest, and the manor house was occupied by the Archbishops as an occasional residence. Harrow church stands on the brow of the hill, and was founded by Lanfranc and consecrated by Anselm in 1094. The present cruciform building of flint and stone has Norman, Early English, and Perpendicular work, and contains some noteworthy brasses dating from the fourteenth century. Harrow School was founded by John Lyon, a yeoman of Preston, which was a hamlet of Harrow. He had a passion for education, and left his entire property on the death of his wife for this school. He also drew up a scheme for the future governance of the school, but it has long outgrown those stipulations and taken its place among the great schools of England. Its masters have almost always been men of mark, and among its scholars are many of our great men. Sir William Jones, Spencer Perceval, Sheridan, Sir Robert Peel, Byron, Lord Palmerston, Lord Shaftesbury, Cardinal Manning, and Lord Lytton are a few of the scholars, poets, and statesmen who were Harrow boys. The school buildings are south of the church and the school-house, built in 1595, is a red-brick and stone Elizabethan structure. The schoolroom, dear to all Harrovians, is a good old room, and its walls are scored with boys' names, not a few of which are dearly prized. "Ducker," a corruption for "Duck Puddle," is the school bathing-place; the name is an ancient one but the bath was reconstructed in its present form in 1881. The handsome school chapel was built by Sir G. Scott in 1857, and the Vaughan Library was begun in 1861, when the first stone was laid by Lord Palmerston. The Library is a noble room, well fitted and furnished, and besides the books, has portraits of Byron, Palmerston, and other illustrious Harrovians. The northern suburb of Harrow is called Greenhill, and stretching towards Harrow Weald is a district known as Wealdstone, a commonplace township. (pp. 10, 16, 45, 46, 50, 51, 57, 63, 92, 95, 100, 136, 137, 143.)

Harrow Weald (2220) is the broad level tract extending from Harrow to Stanmore. As its name implies, it was formerly a wild woodland, and even now it retains some of its old picturesqueness. (pp. 41, 90.)

Hayes (4261) was, in ancient days, a place of some importance, for the manor was held by the Archbishops of Canterbury.

The parish church dates from 1220, and the tower is embattled. The old lych-gate is of interest. There are marble, granite, slate, and brick works in the neighbourhood. (pp. 70, 73, 92, 93, 96, 100.)

Hendon (38,806) is a rapidly growing parish between Hampstead and Edgware. Its name, probably from *Hean-dune*, the high hill, is found in Domesday as *Handone*. The church is on the top of the hill, and the view from the churchyard rivals that from Harrow. The church has a battlemented tower, some interesting monuments, and a Norman font. At the south end of the parish the little river Brent forms a large lake, the Kingsbury Reservoir. Golder's Green and Mill Hill are two hamlets of Hendon, and all around the country is exceedingly pleasant, green, and well-wooded. (pp. 42, 51, 69, 93, 96.)

Heston (15,368), between Southall and Hounslow, was early noted for its fertility, and, while the manor belonged to the Crown, wheat from Heston was used for the royal bread. The church, with the exception of the tower, is modern, and the churchyard is entered by an ancient lych-gate of oak. Osterley House in its fine park is in this parish. Hounslow is partly in this parish and partly in Isleworth. (pp. 68, 73.)

Highgate (10,176), a northern suburb of London, and a ward in Hornsey borough. It commands splendid views of the Metropolis and the surrounding country, and has numerous open spaces, including Highgate Woods. Highgate takes its name from the toll-gate which marked the entrance of the road into the park of the bishop of London. The chapel at the gate-house dates from the time of Edward VI, and the present building has the remains of Coleridge. Highgate has many literary associations, and among its famous residents may be mentioned Andrew Marvell, Bacon, Coleridge, Leigh Hunt, and the Howitts. Highgate School was founded in 1562 by Sir Roger Cholmeley, Knight, Lord Chiet Justice, and re-organised in 1876. The walks in the neighbourhood are still of interest, though the views are being spoilt by the building. The Whittington Almshouses are at the foot of the hill, and opposite is Whittington's stone, now part of a lamp-post. (pp. 40, 45, 46, 50, 57, 61, 79, 130, 134, 136, 138, 140.)

Hillingdon, East (3068) and **West** (7083) are attractive villages on the Uxbridge Road, a short distance to the east of

Highgate School: "Big School"

that town. The parish church of flint and stone has a square tower dating from 1629. There are some good mansions in the neighbourhood, Cedar House and Hillingdon Court being the most famous. (pp. 100, 114.)

Hornsey (84,592), a rapidly-growing suburb of London, was made a borough in 1903. The manor was of great antiquity, and from the earliest times belonged to the Bishop of London, and the name of Bishop's Wood still survives. The old parish church was rebuilt in 1832, but is now disused. Rogers the banker poet was buried in it in 1855. Finsbury Park of about 120 acres is in Hornsey and was opened in 1869. It is laid out in the landscape-garden style, and has some pretty views. Muswell Hill, Fortis Green, and Crouch End are residential districts, and the Alexandra Palace, rebuilt in 1875, has extensive grounds in the borough. (pp. 7, 62, 63, 79, 130, 136.)

Isleworth (27,945), on the Thames, between Brentford and Twickenham, was written originally as Thistleworth. Henry V founded a convent here in 1414, and gave it the name of Syon. The church is partly from Wren's design and has an ancient tower. Isleworth has had many notable residents, and there are some good houses in the neighbourhood, especially Syon House, which stands in grounds stretching for a mile along the river. Isleworth produces large quantities of fruit, and has flour-mills, soap-works, and breweries. The old Borough Road Training College was transferred to Isleworth in 1890. (pp. 26, 66, 69, 70, 80, 91, 102.)

Laleham (478), on the river Thames, two miles south-east of Staines, is a favourite fly-fishing station, having a ferry to Chertsey. This pretty village was the home for some years of Thomas Arnold of Rugby fame; and his son, Matthew Arnold, who was born here in 1822, lies buried in the churchyard. (pp. 10, 22, 140.)

Norwood (26,323), on the west side of Osterley Park is skirted by the Grand Junction Canal. It has a picturesque village green, surrounded by fine trees.

Pinner (7103), on the little brook, the Pin, was once a market-town of importance. Of late years, owing to improved communications, it has developed into a residential resort, and the fields are disappearing. The church, of the fourteenth century, has a fine Perpendicular tower, with battlements and an angle turret. There are many good seats in the neighbourhood, and the Commercial Travellers' School is in this parish. (pp. 39, 50, 51.)

Ruislip (6217) is pleasantly situated between low uplands, watered by two head branches of the Isleworth river, and backed by Ruislip Park, Wood, and Reservoir. There are some picturesque houses and inns in the village, and in the neighbourhood

Pinner Village

are several good mansions. The church stands on high ground, and is a large and interesting building of black flint and stone. (pp. 10, 39, 51, 92, 96, 100, 103, 123.)

Shepperton (2337) is a riverside village and a noted resort for anglers, for the Deeps contain barbel, roach, perch, jack, and trout. The church is a small cruciform Perpendicular building, of which Grocyn, the Greek scholar, was rector in the sixteenth century. (pp. 10, 24, 90.)

Southgate (33,612) was named from its being the south entrance or gate to Enfield Chase. When Charles Lamb lived at Enfield, one of his favourite walks was from Southgate to Colney Hatch. Leigh Hunt was born at Southgate in 1784. (pp. 9, 61, 69, 91, 120, 140.)

Staines (6755), an ancient market-town, stands on the Thames where that river is joined by the Colne. It is on the site of the Roman station *Pontes*, and the name is derived from *stan*,

Stanmore Village

a stone. The London Stone, marking the western limit of the jurisdiction of the City of London over the Thames, still stands in a meadow near Staines Bridge, and bears the inscription, "God preserve the City of London." The present stone, erected in 1812, has also the names of various Lord Mayors and the dates of their official visits. After London Bridge, that of Staines was one of the earliest to span the Thames. The present handsome bridge, designed by Rennie, was opened in 1832 by King

William IV and Queen Adelaide. The parish church has a tower by Inigo Jones, but has been so altered and restored that little of the original building is left. Staines has breweries, mustard-mills, and market-gardens, and a large reservoir for the supply of water to London. (pp. 12, 20, 22, 28, 69, 70, 120.)

Stanmore (**Great**, 1843, and **Little**, 1761). Great Stanmore is one of the highest and most attractive districts in the county. Roman coins, rings, fibulae, pottery, etc. have been found in Stanmore and its vicinity. The present parish church, built in 1849, is the third on this site. The principal house in the district is Bentley Priory, in the midst of a fine and extensive park. Little Stanmore is a quiet agricultural parish lying away from the main road. The chief object of interest is the church of St Lawrence, which was formerly the chapel of the famous palace of Canons. Handel was organist of this church, and the organ on which he played is still in position. (pp. 50, 90, 91, 100.)

Sunbury (4607) is a riverside fishing resort, and here are the rearing-ponds of the Thames Angling Preservation Society. The large reservoirs of the Metropolitan Water Board are in Sunbury, and Kempton Park is famous for its races.

Teddington (17,847), between Twickenham and Hampton Court, was formerly spelt Todington or Totyngton. It has long been a favourite resort of anglers, and is of note with boating men, for it has the first lock on the Thames, 19½ miles from London Bridge. The new lock, opened in 1904, is the largest on the Thames, and is divided into two parts by gates in the centre. The old lock is now used exclusively for pleasure boats. Among Teddington's famous residents may be mentioned Willian Penn, the famous Earl of Leicester, John Walter the founder of the *Times*, and Blackmore the author of *Lorna Doone*. (pp. 7, 10, 20, 21, 25, 70, 88, 140.)

Tottenham (137,418), on the high road to Waltham, is a large working-class town on the Lea. It was formerly a residential

Tottenham High Cross

resort of rich London merchants and Quakers, but there is now little evidence of its former rural character. There is still the High Cross dating from 1609, and the parish church with an embattled tower has some remains of fifteenth century work. The only other building of interest is Bruce Castle, which is named from Robert Bruce, who lived here in Plantagenet times. After undergoing many changes, this red-brick building became a private school in the nineteenth century, and here were educated many pupils who afterwards became famous. Tottenham was formerly noted for its greens, but little more than their names now survive. There are several groups of old almshouses, and other charitable institutions are numerous. There was formerly good fishing in the Lea, which was here crossed by a ferry. The extensive reservoirs and filtering-beds of the Metropolitan Water Board are in the town and neighbourhood. (pp. 7, 62, 63, 69, 70, 129, 130, 143.)

Twickenham (29,367), on the Thames, was noted in the eighteenth century as a place of fashionable resort, and is associated with the names of many distinguished men, especially Pope, who lived for many years at Pope's Villa, where he died in 1744, and Horace Walpole, who built the famous "Gothic Castle" of Strawberry Hill. Twickenham Eyot, better known as Eel Pie Island, is about two acres in extent, and has long been a noted resort of Thames anglers, boat parties, and excursionists. (pp. 7, 10, 21, 66, 70, 85, 86, 103, 131, 132, 135, 137, 138.)

Uxbridge (10,374) is an ancient market-town washed by two branches of the Colne. The parish church, dating from 1447, is not of much interest; but on Lynch Green near by, some protestants were burned in 1555. Uxbridge was the scene of fruitless negotiations between Charles I and the Parliament in 1645, and the house at which the conference was held is known as the Treaty House. Uxbridge has a good corn-exchange; and iron-founding, brick-making, and brewing are carried on, while market-gardening and fruit-growing are important industries. (pp. 39, 69, 70, 81, 91, 122, 130.)

Wembley (10,696) has a fine park, in which is the uncompleted Watkin Tower, which was intended to exceed the Eiffel Tower in Paris by 175 feet. (pp. 51, 70.)

West Drayton (1668), on the Grand Junction Canal, has a fine Perpendicular church of flint and stone. It dates from the early thirteenth century, and the massive tower is unrestored. The font is noteworthy, and there are several brasses and good monuments in the church. (pp. 70, 73, 96, 123.)

Willesden (154,214) is a parish which has shown remarkable growth during recent years. In 1801 the population was 751, in 1841 it had increased to 2957, and in 1881 it had reached 27,363. Since then the improved railway and electric tramway communications have urbanised what was formerly a rural village. The parish church with fragments of Norman work is the only building of historical interest. Willesden is now an important junction on the London and North-Western Railway, having connexions with several other systems. (pp. 39, 62, 63, 128, 134.)

Wood Green (49,369), formerly a hamlet of Tottenham, is now a rapidly-growing district of some importance. The Royal Masonic Boys' School is in Lordship Lane, and there are several other charitable institutions in the place. (p. 7.)

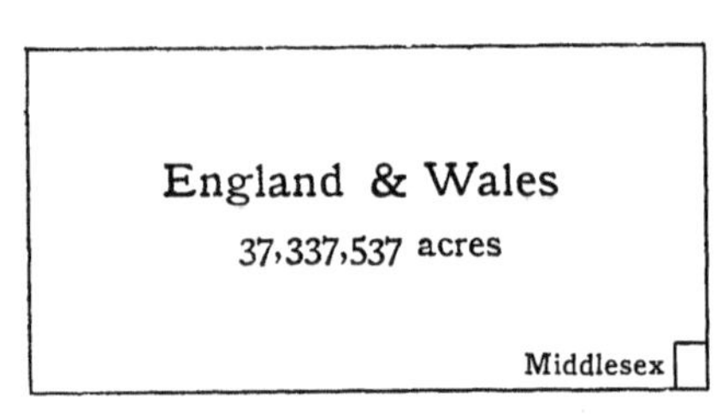

Fig. 1. The Area of Middlesex (148,700 acres) compared with that of England and Wales

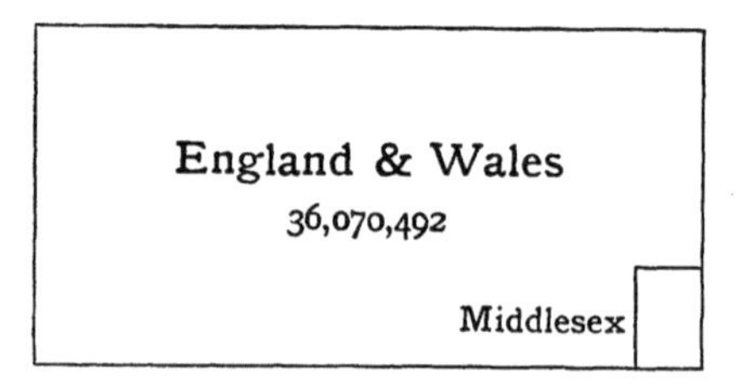

Fig. 2. The Population of Middlesex (1,126,465) compared with that of England and Wales in 1911

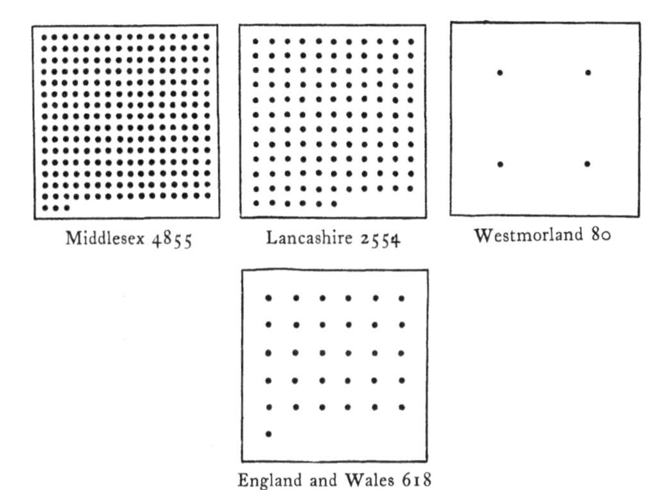

Fig. 3. Comparative density of Population to the square mile

(Each dot represents 20 persons)

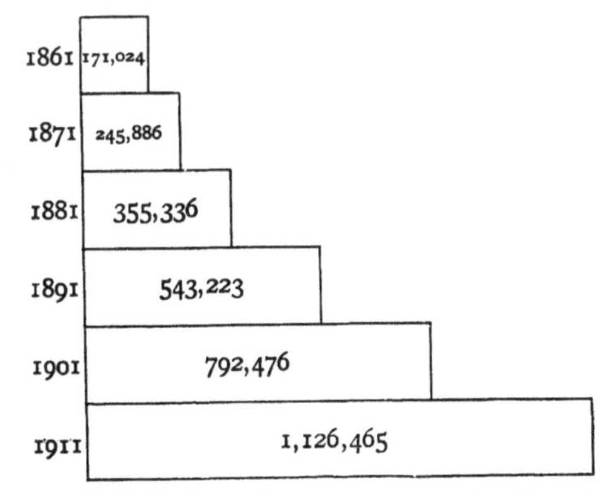

Fig. 4. Diagram showing increase in the Population of Middlesex from 1861 to 1911

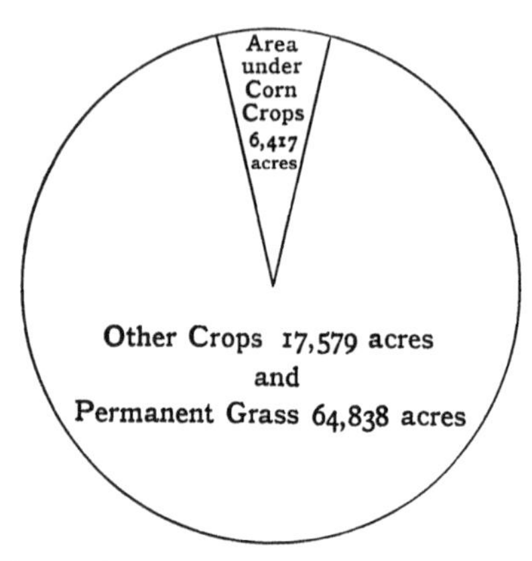

Fig. 5. Proportionate area under Corn Crops compared with other land in Middlesex in 1911

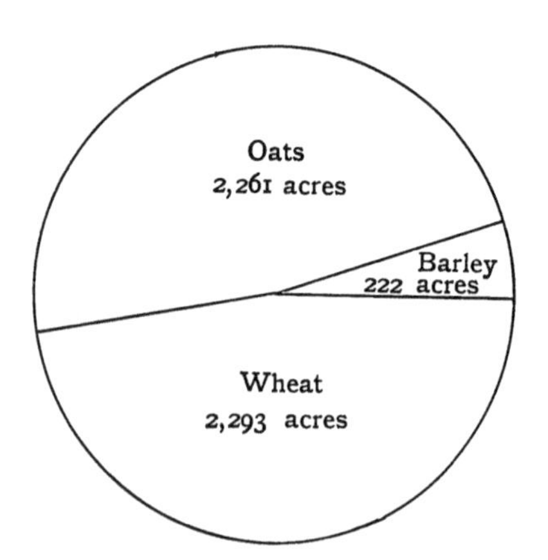

Fig. 6. Proportionate area of chief Cereals in Middlesex in 1911

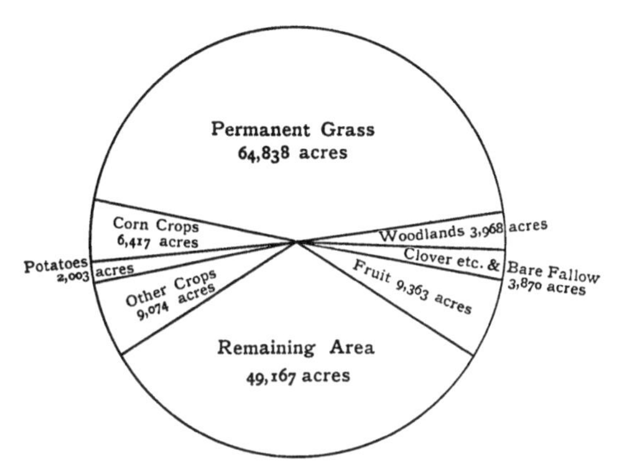

Fig. 7. Proportionate area of Permanent Grass to other areas in Middlesex in 1911

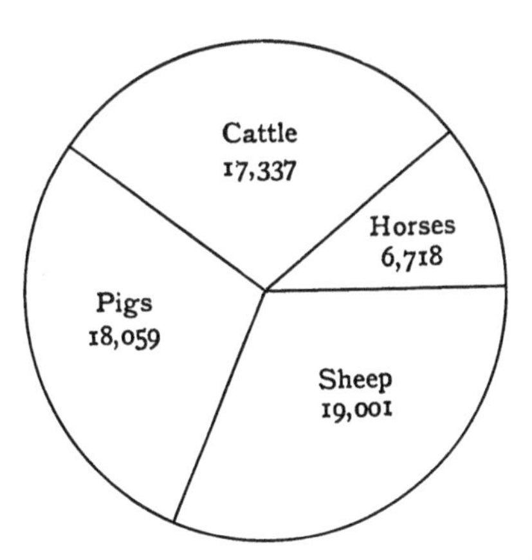

Fig. 8. Proportionate numbers of Live Stock in Middlesex in 1911

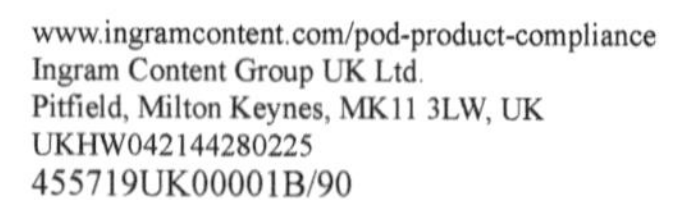

www.ingramcontent.com/pod-product-compliance
Ingram Content Group UK Ltd.
Pitfield, Milton Keynes, MK11 3LW, UK
UKHW042144280225
455719UK00001B/90